Math

WHAT ARE THE ODDS?

CHANCE IN EVERYDAY LIFE

MIKE ORKIN

W. H. FREEMAN AND COMPANY
NEW YORK

Text Design: Diana Blume

Library of Congress Cataloging-in-Publication Data

Orkin, Michael.
　　What are the odds?: chance in everyday life / Mike Orkin.
　　　p.　cm.
　　Includes bibliographical references and index.
　　ISBN 0-7167-3560-1
　　　1. Probabilities—Popular works. 2. Chance—Popular works. I. Title.

QA273.15.O75　1999
519.2—dc21

99-051731

Printed in the United States of America

Second printing, 2000

W. H. Freeman and Company
41 Madison Avenue, New York, NY 10010
Houndmills, Basingstoke　RG21 6XS, England

Contents

Acknowledgments

I'd like to thank the following people at W. H. Freeman and Company, whose help has been invaluable: John Michel, senior trade editor; Sloane Lederer, director of trade sales and marketing; Christopher Miragliotta, project editor; Amy Trask, assistant editor; Diana Blume, designer; and Bonnie Stuart, EPC specialist.

I'd also like to thank Margaret Wimberger-Glickman and Scott Amerman for copyediting my manuscript.

Finally, I'd like to thank my wife, Valerie, whose many creative suggestions added immensely to the shaping of this book.

Introduction

We live in a world of uncertainty and variation, a world that expresses itself in a cosmic mosaic of ever-changing patterns. Some of these patterns may be generated by predictable processes, others by chance. It's often difficult to distinguish between the two. In this book I'll discuss how the laws of chance help describe our world and its patterns and how strange events and bizarre coincidences are part of the natural variation of things. I'll discuss strategies both for games of chance and for everyday interactions and I'll focus on ways to predict long-run results in the face of short-term uncertainty. Among other things, I'll address the following questions:

- If the chance of winning the lottery is so bad, why are there so many winners?
- How come some chance events seem nearly impossible, yet happen frequently?
- How come some chance events are possible, yet never seem to happen?
- If you toss a coin one hundred times, what is the chance of getting heads on every toss?
- Does "survival of the fittest" contradict the notion that chance is a crucial factor in evolution?
- Does randomness really exist, or is the universe a predictable machine?
- How do you analyze a gambling game?
- How do you determine which strategy to use when you have no idea what your opponent will do?
- When should you cooperate, and when should you be nasty?

- How can you be cooperative yet avoid being exploited?
- Is there a dark side to cooperation?
- Can you analyze a war as if it were a game?

From the early cave dwellers to the thousands of modern-day bettors who flock to casinos around the world, humans have always been attracted to games of chance. Games of chance provide metaphors for life's uncertainties. Playing a slot machine or buying a lottery ticket is a symbolic way of handing your destiny to the fates. Understanding the laws of chance can help you decide if some ways of handing your destiny to the fates are better than others.

The laws of chance were originated by seventeenth-century mathematicians trying to find good strategies for gambling games. The results of their research became the foundations of modern probability theory. We will discuss the main principles derived from these efforts and how they can be applied to simple processes, like tossing coins, rolling dice, and casino games. It turns out that powerful tools are available for predicting long-run averages and accounting for chance variation in random processes. Thus, although you can't predict the outcome of a particular bet at the roulette table, you can accurately predict your average winnings (or losses) over many bets.

Processes too complicated for simple analysis can sometimes be modeled with computer simulations. We will look at one such technique, which has been successfully used in situations ranging from the casino game of blackjack to development of the hydrogen bomb. We will also identify the limitations of such techniques: For example, there is no simple game or computer simulation that adequately models international conflicts like NATO versus Yugoslavia.

Outside casinos, chance plays a key role in everyday life. We are uncertain about events in our future that may affect our health, business transactions, or personal interactions. Major turning points in our lives seem to be governed by chance events and random encounters. No matter how hard we try to eliminate uncertainty, chance always seems to sneak through the barricades. Sometimes we get lucky; other times the fates show us that we're not as powerful as we think. Here we will assess some of the tools people use, such as the *I Ching*, to try to predict what fate has to offer.

We will look at the ways in which devices of chance are efficient mechanisms for making unbiased decisions. Chance allows researchers to quickly get accurate information about a large population from a relatively small, random sample. The use of chance helps provide good strategies for certain competitive endeavors. For example, a defending army may be at a disadvantage if the enemy randomly selects a method of attack.

Scientists deal with chance on a regular basis. We will explore how genetics and quantum mechanics substantiate the existence of chance on the molecular and subatomic levels. If there is a cosmic grand plan that determines reality, chance is a key tool for implementing that plan.

I hope that in reading this book you will gain a better understanding of the laws of chance. I hope you will also learn some sensible strategies for surviving in a sometimes hostile, uncertain world.

Chapter 1 ◼

Lotteries, Luck, and Unlikely Events

On April 6, 1999, Maria Grasso, a fifty-four-year-old Chilean immigrant and live-in baby-sitter, won the $197 million Big Game lottery jackpot, the largest lottery prize ever won by an individual. The Big Game ticket had been purchased the same day at a Star Market near Fenway Park in Boston. The winning numbers were 12-17-22-33-44, with "big money ball" number 25. Starting in February, 328 million tickets had been sold in the six Big Game states, but nobody won the twice-a-week drawing from February 2 until April 6.

To purchase a Big Game ticket, you select five "white ball" numbers from 1 to 50, and one "big money ball" number from 1 to 36. In a subsequent Big Game drawing, five winning white balls are randomly drawn from a drum containing balls numbered from 1 to 50, and one winning big money ball is randomly selected from another drum containing balls numbered from 1 to 36. You win the jackpot if your ticket matches all five winning white balls, in any order, and the big money ball. There are 76,275,360 Big Game ticket combinations. Thus, the chance of winning the Big Game jackpot is about 1 in 76 million.

Why did Maria Grasso win the jackpot? Luck is a group phenomenon. There were approximately 83 million tickets sold for the April 6 drawing. Maria Grasso was the one who got lucky.

The biggest lottery in America is Powerball, currently played in twenty states and Washington, D.C. On Wednesday, July 29, 1998, the Powerball jackpot prize reached $295.7 million, the largest single lottery prize in the history of the world. The winning ticket had the numbers 8, 39, 43, 45, 49, with Powerball number 13, and was purchased by thirteen machine-shop workers from Ohio who pooled their money to buy 130 Powerball tickets at a Speedway gas station and convenience store in Richmond, Indiana. The Powerball jackpot prize of $295.7 million could have been paid in yearly installments over twenty-five years or as a lump sum. The winners chose the lump-sum option of $161.5 million.

In addition to the winning jackpot ticket, there were 210,850,581 losing tickets in the July 29 drawing. In the nineteen drawings with no jackpot winner that led to the July 29 drawing, there were 562,434,262 losing tickets. Again, luck is a group phenomenon: For every lucky jackpot winner there is a multitude of losers.

To purchase a Powerball ticket, you pick five "white" numbers from 1 through 49 and one Powerball number from 1 through 42. In a subsequent Powerball drawing, five winning white balls are randomly drawn from a drum containing the numbers 1 to 49, and one winning red ball, called the Powerball, is randomly drawn from a drum containing balls 1 through 42. There are nine prize categories altogether. As with the Big Game, the category of interest is the jackpot, which is won when your ticket matches all five winning white balls, in any order, and the Powerball. There are 80,089,128 possible Powerball ticket combinations, so the chance of winning the Powerball jackpot is about 1 in 80 million. Here are some analogies to illustrate what the Powerball odds really mean (the Big Game is not much different):

- If you buy 50 Powerball tickets a week, you will win the jackpot on the average of about once every 30,000 years.

- If you fill 1,200 Soldier Fields (Soldier Field is the home of the Chicago Bears) and give everyone in the 1,200 Soldier Fields a Powerball ticket, there will be an average of one jackpot winner.

- Suppose that every time you drive a mile you buy a Powerball ticket. Then you will have to drive an average distance equal to 167 round trips to the moon before you win the jackpot.

Lotteries and Lightning

It is fashionable to compare your chance of winning the lottery to your chance of getting struck by lightning. If you stay in your house during a thunderstorm, however, the chance is essentially zero that you will get struck by lightning, whereas if you stand in the middle of a golf course, the chance is considerably greater. If you live in an area with few thunderstorms, like San Francisco, your chance of getting struck by lightning is very small, indeed. Thus, unlike lottery drawings, which are completely random (every ticket has the same chance of being a winner), the chance of getting struck by lightning indicates only what fraction of the population has been struck over a given period, not how likely it is to happen to you. The same is true for other popular "odds," such as the chance of having a heart attack or becoming a raving lunatic.

Lottery Strategies

If you buy a Big Game ticket, the chance is about 1 in 76 million that you will win the jackpot. If you buy a Powerball ticket, the chance is about 1 in 80 million that you will win the jackpot. Gamblers aren't happy with such bad odds and develop strategies, like the following, to improve their situation:

When the jackpot gets large, buy every possible ticket. When the Big Game lottery reached the stratospheric jackpot of $197 million, a desperate player might have considered the strategy of investing $76 million and purchasing every possible ticket combination, thus guaranteeing a jackpot win and, in this case, a big profit. Purchasing every ticket combination when the jackpot gets high is risky business, because multiple winners split the pot. For the April 6 drawing, about 83 million tickets were purchased, but there was only one winner, Maria Grasso, who decided to take the jackpot in a lump sum of $104 million. If you had invested the approximately $76 million necessary to buy every possible ticket, you would have had to split the prize with Maria Grasso, yielding you a lump sum of about $52 million, plus your share of the part of your ticket investment that went into the jackpot pool. You would have ended up with a loss of a few million dollars. There is also a logistical problem with purchasing every ticket. If you fill out three tickets per minute, twenty-four hours per day, it would take you about fifty years to write down all 76 million combinations.

Buy lots of tickets. Many lottery players purchase more than one ticket. For example, the thirteen Powerball players whose winning ticket won the big jackpot bought 130 tickets. The theory is this: The more tickets you buy, the greater your chances of winning. The theory is true—but it's also true that the more tickets you buy, the more money you spend. Your average payoff doesn't change.

Bet on numbers that are due. A popular lottery strategy is to bet on numbers that haven't come up for a while—in gambling jargon, numbers that are due. The theory is that numbers that haven't been selected for some time are more likely to come up than ones that have been selected. The theory is false. Although the fraction of occurrences of each lottery number will be about the same in the long run, chance will cause bizarre streaks and weird patterns. If the lottery's random-selection device is operating properly, all numbers have the same chance of being selected. Lottery numbers are never due.

Bet on numbers that are hot. Another popular strategy is to bet on "hot" numbers, that is, numbers that have come up more frequently than chance would seem to allow. Actually, if for some reason the random-selection equipment isn't working properly and is biased for particular numbers, such numbers will indeed be hot. A clever lottery player could take advantage of this situation and make a killing. Unfortunately for gamblers, lottery equipment is well maintained and usually works properly. In addition, amazing streaks of seemingly hot numbers occur by chance and make real biases hard to detect.

Since no popular lottery strategies work, why buy lottery tickets at all? Here are some reasons to play:

- For a small investment, lotteries give you a chance, albeit a miniscule one, to dramatically change your lifestyle.
- Buying lottery tickets is a social event that encourages you and your friends to pool your resources and work toward a common goal. Consider the thirteen machine shop workers who pooled their resources and won the huge Powerball jackpot. Also consider the millions of others who pooled their resources and lost.
- Unlike with the stock market, with the lottery, if you buy tickets, you don't have to pay a broker's fee.
- This could be your lucky day.
- Bingo doesn't offer such big jackpots.
- Lotteries are a good mechanism for the random distribution of wealth.
- Buying lottery tickets is better for your health than spending the money on cigarettes.

The Lottery Principle

With such astronomical odds against winning the Big Game and Powerball, why are there usually winners? This is due to what I call the lottery principle (statisticians Persi Diaconis and Frederick Mosteller call it the law of very large numbers): *Given enough opportunity, weird things will happen just due to chance.*

There are jackpot winners because there are so many players—millions of tickets are bought for each drawing. Since the chance of winning the Big Game jackpot is about 1 in 76 million, having 1 winner out of 83 million tickets sold for the April 6 Big Game drawing conforms to the statistical average. For the July 29 Powerball drawing, 210,850,582 tickets were purchased. The surprising thing wasn't that there was a winning ticket but that there was only one winner. With the chance of winning equal to 1 in 80 million, there is a theoretical average of 2.6 winners when 210 million tickets are purchased. To have only one winner is somewhat less than this average.

One thing is clear: Luck is a group phenomenon. For every Big Game winner, there are about 76 million losers. For every Powerball winner, there are about 80 million losers. When you buy lottery tickets, however, don't worry about your terrible chances of winning. Instead, think of the big picture. You belong to the group of players. If the group is large enough, someone will win. Even though the someone probably won't be you, your participation adds to the chance that that there will be a winner.

The lottery principle applies to seemingly weird, magic, and mystical occurrences. Whenever a bizarre event takes place, it may be merely a random fluke supported by a banal background of boring behavior. Only the weird is noticed. The ordinary is ignored. Here are more examples of the lottery principle:

Economic Forecasters

Every year for ten years, 1,000 economic forecasters each toss a coin to predict whether the economy will go up or down. There is likely to be a lucky forecaster with a perfect ten-year record due to chance alone. The newly anointed expert is the one who got lucky. The other 999 forecasters were predicting the economy exactly like the new expert, however, they were the unlucky ones.

Note: If you always toss a coin to predict whether or not *anything* will happen, you will be right about half the time.

Fortune Cookies

The message in Mr. Spiff's fortune cookie says, "You will soon be hit by a bus." Disregarding this dire prediction, Mr. Spiff goes to work as usual the next morning and, while crossing a downtown street, gets hit by a bus as it pulls away from the curb. Fortunately, Spiff recovers. He vows that he will never again ignore a fortune cookie's prophecy. What he doesn't know is that there were half a million other cookies with the same prophecy. Although it is unlikely that a particular person will be hit by a bus soon after a fortune cookie predicts it, with half a million such predictions, it is likely to happen to someone.

Predicting Earthquakes

The noted psychic, Madam Zola, has correctly predicted every major earthquake in California for the past twenty years. What most people don't know is that Madam Zola predicts a major earthquake every day. People don't take notice of her numerous incorrect predictions, but on the rare occasions when there are major earthquakes, everybody remembers Madam Zola.

Bombing Yugoslavia

Suppose that NATO forces drop 100 "smart bombs" each day on military targets in heavily populated areas of Yugoslavia. Suppose there is a 99.5 percent chance that a smart bomb will not cause civilian casualties. Then each month, just by chance, there will be an average of 15 smart bombs that cause civilian casualties. In a given week there is a 97 percent chance that at least one smart bomb will accidentally cause civilian casualties.

Given enough smart bombs dropped on military targets in populated areas, some will cause civilian casualties no matter how "smart" they are. "Dumb," unguided bombs are considerably more haphazard.

Nuclear Accidents

The chances may be small that there will be an accident at a particular nuclear power plant, but with many plants, over a period of time, accidents will happen as surely as there are lottery winners. Suppose there is a 99.9 percent chance that a particular power plant is safe and won't have a serious accident in a given year. If there are 300 such plants in operation, there is a 95 percent chance that there will be at least one serious accident every ten years, with an average of three serious accidents every ten-year period. Given enough opportunity, unlikely accidents will occur. If the unlikely accident will cause environmental havoc, "99.9 percent safe" could spell eventual disaster.

Knowing the chance of a system failure is useless unless you also consider the number of systems in operation and the amount of damage a failure will cause.

Pollution

Suppose the chance is of an individual getting cancer from prolonged exposure to a certain pollutant is 1 in 1,000. If this pollutant is present in the atmosphere of a city with 1 million inhabitants, about 250 of them will get cancer.

A low chance of an adverse side effect should be considered in the context of the number of people exposed and the harmfulness of the side effect.

Evolution

The lottery principle sheds an interesting light on evolution. If there are a variety of organisms in a friendly environment, some lucky organisms will be successful. Breeding and mutation provide mechanisms for genetic mixing. Resulting diversity affords an opportunity for survival that would not otherwise exist.

In the genetic lottery, survival is a group phenomenon: For every winning species, there are many losers who become extinct. Maybe survival of the fittest should be changed to *survival of the luckiest.*

Horatio Alger Stories

One of the moral underpinnings of our society is the work ethic: Work hard and you will succeed. The heroes of the nineteenth-century rags-to-riches stories by Horatio Alger personified this ethic. A part of the story that isn't popularized is the statistical background: For every lucky winner who starts poor, works hard, and becomes a success, there are many more losers who start poor, work hard, and stay poor. Even if the chance of going from rags to riches is small, the lottery principle insures that if enough people try, a few will succeed. I suggest a modification of a popular rallying cry. Instead of "Work hard and you will succeed" I propose this: "Work hard, and your chances of success are greater than if you don't work hard, unless you have a large inheritance, in which case it doesn't matter what you do, as long as it's not something really ridiculous."

But perhaps that's too cynical. Here's another way to look at it: "Work hard and don't be put off by failure. Even if it is unlikely that you succeed on any particular endeavor, the lottery principle insures that if you try enough times, you will eventually succeed."

Stockbrokers

A clever stockbroker advertises that you only pay a broker's fee for her correct decisions on whether to take long or short positions on selected stocks; if her decision is wrong, you owe her nothing. Unbeknownst to her clients, the broker tosses coins to make her decisions. Regardless of the ups and downs of the market, she will be correct about half the time just by chance. Given enough customers, the broker will make a handsome profit. The losers won't have to pay a broker's fee for the bad advice, but they may lose serious money from their investments.

The Law of Nonoccurrence

Some things are so unlikely that there simply isn't enough opportunity for them to occur. In this case the lottery principle, although true in theory, fails in practice. I call this phenomenon the law of nonoccurrence. Here are some examples:

The Really Bad Lottery

Suppose that there is a lottery in which you bet on twenty numbers between 1 and 80. In the subsequent drawing, twenty winning numbers are drawn at random. You win the jackpot if all twenty of the numbers you bet on match the twenty winning numbers. The chance of winning this jackpot is less than 1 in a quintillion (1,000,000,000,000,000,000). If there is one drawing per week and everyone on earth (6 billion people) always buys a ticket, it will take an average of about 5 million years to produce a winner.

The Flip Side of Coin Tossing

If you toss a coin 100 times, you'll probably get somewhere between 40 and 60 heads. What chance do you have of getting 100 heads in 100 tosses? Let's informally compute the probability. If you toss a coin once, the chance of getting heads equals 1/2. If you toss a coin twice, the chance of getting heads both times equals $1/2 \times 1/2 = 1/4$. The chance of getting 3 heads in 3 tosses equals $1/2 \times 1/2 \times 1/2 = 1/8$, and so on. The chance of getting 100 heads in 100 tosses equals 1/2 to the 100th power, or 1 in 1,000,000,000,000,000,000,000,000,000,000, 000,000. That's 30 zeros, my friend. If every person on earth (6 billion people) starts tossing coins 24 hours per day, with each person tossing at the rate of 100 tosses every 5 minutes, it will take an average of about a million billion (that's 1,000,000,000,000,000) years until somebody gets 100 heads in 100 tosses. Even though it's theoretically possible, getting 100 heads in 100 tosses is, for all practical purposes, impossible. If the 6 billion people now on earth had been doing this since the beginning of the universe, it's unlikely that anyone would have yet come up with 100 heads in 100 tosses. Compared with this, winning the Powerball jackpot is a piece of cake.

Salt, Pepper, and the Second Law of Thermodynamics

The laws of thermodynamics were developed by nineteenth-century scientists to describe relationships between heat and mechanical energy.

The second law of thermodynamics is based on the observation that many physical processes appear irreversible: Although these processes may theoretically be able to proceed forward or backward, when we observe them, they only go one way. Once a supposedly irreversible event occurs, there's no turning back. The general version of the second law of thermodynamics, informally stated as follows, addresses this phenomenon: *The total perceived disorder, or "entropy," in the universe is always increasing.*

Suppose you put some salt in a container and put a layer of pepper on top of the salt. If you shake the container vigorously, the salt and pepper become disorderly, or randomly mixed. Suppose you keep shaking the container in an attempt to get the mixture to return to its original state, with the salt separated from the pepper. It is theoretically possible that you will get lucky and succeed, but this is so unlikely that not even the most degenerate gambler would bet on your success. The reason is simple: Even though every configuration of salt and pepper in the container has the same chance of occurring, there are so many more configurations of salt and pepper grains in which the ingredients are mixed together than configurations in which they are separated that the chance of separation by random mixing is near zero. It's like trying to get 100 heads in 100 coin tosses: There are so many more heads-tails combinations that yield a result close to 50 heads and 50 tails that, even though it's possible to get 100 heads in 100 tosses, it won't happen. Similarly, you can shake the container until the cows come home, but you'll never separate the salt and pepper.

The salt-and-pepper example suggests that, in part, the second law of thermodynamics has a statistical explanation: *There are more ways for disorder to occur than order.*

Just as electricity follows the path of least resistance, chance follows the path of most likely. Although it is theoretically possible for things in the universe to get more orderly, they eventually get disorderly, simply because disorder can happen in more ways than order.

Now think of Humpty Dumpty. Suppose you push an object off a wall so that it falls to the ground and smashes to pieces. Maybe it's possible to meticulously put all the pieces together again, but it requires a lot of work. This suggests another factor contributing to the second law of thermodynamics: It's easier to destroy something than to reconstruct it later.

It may even be theoretically possible to put a broken object together again by some form of random mixing; however, like separating the salt from the pepper, there are just too many configurations of the pieces for random mixing to yield the original object.

Salt and pepper and Humpty Dumpty show how chance is intertwined with the second law of thermodynamics: Because of the lottery principle, given enough opportunity, weird events, like objects falling off walls, will happen just due to chance. Because of the many combinations that the resulting messes can assume, things don't always get put back together again.

The Law of Absurdity and the Random Dart

Let's return to the coin. In addition to the near impossibility of getting 100 heads in 100 tosses, it is equally impossible to get any other *specific sequence* of heads and tails in 100 tosses. For example, it's nearly impossible for 100 tosses to result in alternating heads, tails, heads, tails, or to get heads on the first 25 tosses, tails on the next 25 tosses, heads on third 25, and so on. Or to have the first 50 tosses land heads and the second 50 tosses land tails. In fact, there are so many heads-tails combinations in 100 tosses that the chance of any *particular* sequence is nearly zero. The same is true for the randomly mixed salt and pepper: There are so many different salt and pepper configurations in which the precise location of every grain of salt and pepper is specified that the chance of any particular configuration is essentially zero. These situations are examples of a phenomenon that I call the law of absurdity: *Everything is impossible, yet something must happen.*

The law of absurdity seems to contradict the law of nonoccurrence, but these laws are really just two manifestations of the same phenomenon. The law of nonoccurrence says that any particular, extremely unlikely event most likely won't occur, while the law of absurdity says that in a random selection from a very large collection of unlikely events, *some* event (not any particular event) will occur. Here's another example.

Harkness has drunk one too many flagons of ale; however, if you point him in the direction of a very large dartboard, he will toss a dart that is guaranteed to hit the board but will do so at random in such a way that every point on the board has an equal chance of being hit. Thus, there's a 50 percent chance that the random dart will hit the lower half of the board, a 25 percent chance that the dart will hit the upper right quadrant, and a 5 percent chance that the dart will hit the small circle surrounding the bull's-eye that takes up 5 percent of the board's area. In general, to find the chance that the random dart will hit any region on the board, you take the area of the region and divide by the area of the board.

What is the chance that Harkness's random dart will hit a particular point on the dartboard?

Even the tip of a very sharp dart occupies a small amount of area. (If you look at a sharp dart under a strong microscope, it doesn't look so sharp.) So we must compute the chance that the dart will hit a particular tiny region, equal in area to the tip's cross-section, around the specified point on the dartboard. The sharper the dart, the smaller the region, and so the chance that Harkness's sharp, random dart will hit a particular point on a large dartboard is close to zero. Yet the dart has to hit somewhere. I call this phenomenon *the dart of Harkness.*

The Random Trout

A fish scientist predicts that a certain trout will have a length between 7 and 11 inches at maturity, with its precise length randomly distributed in this range. (For simplicity, let's assume that the trout can't be less than 7 inches or more than 11 inches long). What is the chance that the trout will be *exactly* 9 inches long? The situation is analogous to the dart of Harkness: To find the chance that the trout length will be in any particular interval in the 4-inch range between 7 and 11 inches, divide the length of that interval by 4. For example, the chance that the trout will be between 8 and 10 inches long equals 2/4 = .5. The chance that the trout will be between 8.5 and 9.5 inches long equals .25. The chance that the trout will be between 8.8 and 9.2 inches long equals .1. The chance the trout will be exactly 9 inches long is less than the chance that its length will be in any small interval that contains 9 inches. Since we can make such intervals as small as we like, it follows that the chance that the trout is exactly 9 inches long (or any other exact length) equals zero. Yet, the trout has to be *some* length. The law of absurdity rears its smiley face again, this time in the context of measurement precision.

The confusion lies in what we really mean when we say that a trout is 9 inches long (or that a person is 6 feet tall, or that a dog weighs 50 pounds, or any other physical measurement). In fact, what we really mean is that our measurement is within some small increment of the convenient, exact amount, with this increment depending on the precision of our measuring device. Although there are theoretically an infinite number of possible trout lengths between 7 and 11 inches, rulers and other measuring devices can only be calibrated to a finite level of precision. If the fish scientist uses a ruler calibrated to sixteenths of an inch, there are only $16 \times 4 = 64$ possible measured lengths in the 7-to-11-inch range. Using such a ruler, when we say that a trout is 9 inches long, we mean that its length is in the sixteenth-of-an-inch range that is closer to the 9-inch mark on the ruler than the 8-and-15/16-inch mark and the 9-and-1/16-inch mark. According to this particular ruler, the probability that a randomly selected trout will be 9 inches long is 1/16 divided by 4, or 1/64. If the ruler is calibrated to thirty-seconds of an inch, the chance that a randomly selected trout will be 9 inches long is 1/128, and so on. Here, probabilities depend on the precision of the measuring device. The most precise that the words "9 inches long" can ever be is some small increment *around* 9 inches.

At a certain microscopic level, a trout's length is an uncertain quantity, regardless of the measuring device. Dirt falls off, algae clings to its surface. On this level, the trout, as well as everything else in the universe, is in a constant state of change. Thus, in an absolute sense, it's impossible for a trout (or anything else) to be exactly 9 inches long.

Chapter 2 🎲

The Laws of Chance

Probability

People have been using chance devices since cave dwellers fashioned dice from animal bones to play games and get messages from the gods. Modern probability theory was developed in the seventeenth century when gamblers consulted mathematicians to find winning strategies. The laws of probability quantify chance by estimating the likelihood that an event will occur. The probability of an event should approximate the fraction of times the event will occur in many repetitions of an experiment or process.

Probabilities are expressed as percentages or fractions; we say there is a 50 percent chance that a coin comes up heads, or the chance is one-sixth that 1 comes up when you roll a die. Here is a problem that motivated French mathematicians to study the laws of chance.

The Chevalier de Mére's Dice Games

In 1654 French nobleman and saloonkeeper the Chevalier de Mére had a problem with a dice bet. Actually, two bets. In the first bet the Chevalier gave even-money odds (winners win the amount they bet) that in 4 rolls of a die, at least one 6 would come up. If no 6's came up, the Chevalier lost the bet. He correctly reasoned that in one roll, the chance equaled 1/6 that 6 would come up. He incorrectly reasoned that in 4 rolls, the chance was 4/6, or 2/3, that at least one 6 would come up. The correct chance of getting at least one 6 in four rolls, rounded off to two decimal places, equals .52. Even though his reasoning was wrong, the Chevalier had the best of it—a winning proposition—and made a profit in repeated play.

When the betting dropped off, the Chevalier offered a second bet: even-money odds that in 24 rolls of a pair of dice, at least one double 6 would come up. He correctly reasoned that since there are 6 faces on one die, there are $6 \times 6 = 36$ combinations of upturned faces with a pair of dice, and so in one roll of a pair of dice, the chance equals 1/36 that a double 6 will come up. The Chevalier incorrectly reasoned that in 24 rolls, the chance is 24/36, or 2/3, that at least one double 6 will come up. The correct chance of getting at least one double 6 in 24 rolls, rounded off to two decimal places, equals .49. Bad news for the Chevalier, who now had a losing proposition. In despair, he consulted mathematician Blaisé Pascal, who, after corresponding with fellow mathematician Pierre de Fermat, discovered what was wrong with Chevalier's reasoning and started laying the foundations of modern probability theory.

It turns out that if the Chevalier had allowed 25 rolls instead of 24 to get a double 6 in his second wager, he would have had a winning proposition instead of a losing one. He wouldn't have needed to consult with Pascal, possibly delaying the development of probability theory.

In 1952, a New York gambler known as Fat the Butch gave even-money odds that in 21 rolls of a pair of dice he would get at least one double-6. Apparently he hadn't heard of the Chevalier. In a series of bets with a gambler known as The Brain, Fat the Butch lost $50,000.

Basic Probability Calculations

When every possible outcome of an experiment has the same chance of occurring, there is a simple method for calculating probabilities. This method can be applied to tossing coins, rolling dice, dealing cards, drawing lottery numbers, random sampling, and many other processes. In such situations, to find the probability of an event, simply count the number of ways the event can happen and divide by the total number of possible outcomes. Here are some examples:

Tossing Coins

When you toss a coin, it can come up either heads or tails. The probability that the coin comes up heads equals 50 percent or 1/2. Same for tails.

Rolling Dice

A die has 6 faces, numbered from 1 to 6. The probability that 1 comes up equals 1/6. The probability that 3 comes up also equals 1/6.

Dealing Cards

There are 4 aces in a deck of 52 cards. When you deal a card from well-shuffled deck, the probability that an ace is selected equals 4/52 = 1/13. The probability that a heart is selected equals 13/52 = 1/4. The probability that the ace of hearts is selected equals 1/52. The probability that an ace or a heart is selected equals 16/52.

Playing Lotteries

Draw a ball at random from a box containing balls numbered from 1 through 51. The probability that you select number 17 equals 1/51. In the California Super Lotto game, 6 balls are randomly selected from a box containing balls numbered from 1 through 51. There are about 18 million combinations of 6 numbers drawn from 51, so if you buy a ticket to the Super Lotto game, your chance of winning equals 1 in 18 million. In other words, the odds against winning California Super Lotto are about 18 million to 1.

Box Models

Lotteries are just one example of a box model, the basis of which is putting balls in a box and selecting some at random. Many chance processes can be represented as box models.

If you put 52 balls in a box, labeled as cards in a deck of playing cards, then drawing 5 balls at random is the same as dealing a poker hand.

If you put 2 balls in a box, one marked heads and the other marked tails, drawing a ball at random is the same as tossing a coin. Drawing a ball at random, putting the selected ball back in the box, drawing another ball, putting it back, doing this 20 times is the same as tossing a coin 20 times.

If you put 6 balls numbered from 1 through 6 in a box and then draw a ball at random, you are, in effect, rolling a die. Drawing a ball, putting it back in the box, and drawing another ball is the same as rolling a pair of dice.

Combinations

Finding probabilities can be tedious because of the many ways things can occur. Suppose you toss a coin 4 times. Since there are 2 possible outcomes for each toss, there are a total of $2 \times 2 \times 2 \times 2 = 16$ possible combinations of heads and tails in the 4 tosses: HHHH, HHHT, HHTH, HTHH, THHH, HHTT, HTHT, THHT, THTH, TTHH, HTTH, HTTT, THTT, TTHT, TTTH, TTTT.

Of the 16 heads-tails combinations, there are 6 that make up the event "2 heads and 2 tails": HHTT, HTHT, THHT, THTH, TTHH, HTTH. Thus, the probability of getting 2 heads and 2 tails in 4 tosses equals $6/16 = 3/8 = 37.5\%$.

If you toss the coin 10 times, things start to get ugly. There are 2 to the 10th power = 1,024 heads-tails combinations for 10 tosses, too many to conveniently put in a list. How many heads-tails combinations yield 5 heads and 5 tails? Problems like this necessitated the development of a combinations formula, which was accomplished by seventeenth-century mathematicians. It turns out that there are 252 combinations of 5 heads and 5 tails, so the probability of getting 5 heads and 5 tails in 10 tosses equals $252/1024 = 24.6\%$. Since there is only one coin-tossing sequence that yields heads on every toss, the probability of getting 10 heads in 10 tosses equals $1/1024$, or approximately 1/10 of 1%. In 100 coin tosses, there are an astronomical number of heads-tails combinations. There are also an astronomical number of heads-tails combinations that yield 50 heads and 50 tails. When the smoke clears, the probability of getting 50 heads and 50 tails in 100 tosses, allowing for any order of occurrence, equals 7.96 percent, or about once in every 12.5 attempts.

The Law of Averages

If you toss a coin, the probability that heads comes up equals .5. If you roll a die, the probability that 1 comes up equals 1/6. A card dealt from a well-shuffled deck has a 1/13 chance of being an ace. What do these numbers really mean? It turns out that for repeated results of certain processes, such probabilities accurately predict the long-run fraction of occurrences of an event. This is made precise by the law of averages, which, informally stated, means that *in many independent trials of an experiment, the observed fraction of occurrences of an event eventually gets close to the probability of occurrence on one trial.* An independent trial of an experiment is one that is unaffected by the results of other trials. Examples of independent trials are coin tosses, dice rolls, roulette spins, and lottery drawings.

Did you hear about the statistician who took a bomb with him whenever he got on an airplane? He did this as a safeguard against terrorism, reasoning that although the chance is low that a terrorist will bring a bomb onto a particular airplane, the chance is *really* low that *two people* will bring bombs onto the same plane. Actually, whether a terrorist brings a bomb onto an airplane is *independent* of what the statistician does, and so the chance of the terrorist bringing the bomb on board remains the same.

Probability provides precise predictions. The law of averages gives meaning to statements like "The probability is .5 that a coin comes up heads" because it guarantees that in repeated coin tosses the observed fraction of heads will eventually get extremely close to .5, no matter what weird patterns occur along the way. For example, eventually the fraction of heads will be between .4999 and .5001 and stay that close to .5 forever. You specify the precision, and eventually you get it. The meaning of *eventually* depends on the level of precision you want. Thus, although you may not be able to predict the outcome of a particular coin toss, you can make accurate predictions of long-run, average results.

Computer Simulations

To demonstrate the law of averages, I wrote a computer program that simulated 72,000 rolls of a pair of dice. Computer dice rolls aren't quite the same as real dice rolls, because computers use a formula that generates numbers that *appear* random but that can be reproduced if the starting value of the sequence is known. This is helpful for scientists who need to reproduce experiments but an anathema for randomness purists. For practical purposes, computer-generated dice rolls are fine. Also, it's much quicker to use a computer than to actually roll dice. If you manually roll dice at the rate of 10 rolls per minute, 24 hours per day, it will take 5 days to roll a pair of dice 72,000 times. My computer does it in 3 seconds.

Since there are 6 faces on one die, there are $6 \times 6 = 36$ possible outcomes for a pair of dice. I've listed these outcomes as follows, with the left number denoting the result of one die (call it the yellow die) and the right number denoting the result of the other (call it the blue die):

36 dice outcomes (y = yellow die, b = blue die)

y,b	y,b	y,b	y,b	y,b	y,b
1,1	1,2	1,3	1,4	1,5	1,6
2,1	2,2	2,3	2,4	2,5	2,6
3,1	3,2	3,3	3,4	3,5	3,6
4,1	4,2	4,3	4,4	4,5	4,6
5,1	5,2	5,3	5,4	5,5	5,6
6,1	6,2	6,3	6,4	6,5	6,6

There are 6 outcomes that sum to 7: 1,6 6,1 2,5 5,2 3,4 4,3.

Thus, when you roll a pair of dice, the probability that 7 comes up equals $6/36 = 1/6 = .16667$, rounded off to 5 decimal places. The law of averages states that in a large number of dice rolls, the fraction of 7's should be close to .16667. In 72,000 rolls, the theoretical average number of 7's is 1/6 of $72,000 = 12,000$. I ran the program 10 times and got the following results:

**10 repetitions of
72,000 computer-simulated dice rolls**

fraction of 7's	number of 7's
.16872	12,148
.16742	12,054
.16667	12,000
.16664	11,998
.16383	11,796
.16596	11,949
.16542	11,910
.16729	12,045
.16692	12,018
.16564	11,926
average: .16645	11,984.4

The simulated results varied, ranging from a low of 11,796 occurrences of 7 (.16383) to a high of 12,148 occurrences (.16872), with an average of 11,984.4 occurrences of 7 per 72,000 rolls (.16645). The law of averages states only that the long-run fraction of 7's will be approximately .16667. The results of the dice-roll simulations were in the normal range of chance variation.

Pure Random-Number Generators

According to an article in *PC Magazine* (June 1999), Intel has developed an inexpensive computer random-number generator that uses the "thermal noise" generated by heat variation coming from a computer chip's resistor. Unlike the traditional formulas that generate deterministic, *pseudo-random* sequences (unpredictable only if you don't know the formula), thermal noise is considered a source of *pure randomness* (the jury is still out on this). That, coupled with traditional methods, will increase the effectiveness of computer-generated random numbers. Other methods of generating pure random numbers from hardware have been developed, but, because of cost and size factors, they haven't been mass-produced as standard components on desktop computers.

Chance Variation

The law of averages asserts that in many dice rolls, the fraction of 7's will be approximately .16667. The computer simulation seemed to verify this, but *approximately* is a term that begs to be quantified. The reason for using terms like *approximately* or *close to* is because of chance variation, an inherent part of randomness. The law of averages is a limit theorem: It states precisely what will happen in an infinite number of observations, something we will never see. When we talk about the long run in practical terms, we mean a large but finite number of observations, and so we must allow for chance variation. Chance variation is best understood in the context of how things vary around the average value.

Average Number of Occurrences

If you toss a coin 10 times, it's possible to get any total between 0 and 10 heads. Since the chance of heads on any particular toss equals 1/2, the *theoretical average* in 10 tosses equals $10 \times 1/2 = 5$ heads. Sometimes you'll get more and sometimes less, but in the long run things average out to about 5 heads in every 10 tosses. In general, the average number of occurrences equals the chance of an occurrence on one trial times the total number of trials. In 72,000 rolls of a pair of dice the average number of occurrences of "7" equals $72,000 \times 1/6 = 12,000$. The law of averages says that in a large number of trials, the actual number of occurrences will be close to the average number of occurrences, with some variation due to chance.

Averages Don't Tell the Entire Story

Did you hear about the 6-foot-tall person who drowned in a lake whose average depth was 2 feet?

Again, knowing only the *average* isn't a complete description of a process or system. We must also know something about *variation*. Here is why the 6-footer drowned in the shallow lake: Even though the *average* depth was 2 feet, there was considerable *depth variation*. The drowning took place in a hole that was 10 feet deep. (A shallow pool on the other side of the lake compensated for the deep hole.) This lake, with variable depth, is quite different from a flat-bottomed lake with a uniform depth of 2 feet, even though the average depths are the same. To have a better picture of the lake, we must know how specific depths vary from the average.

Standard Deviation

Chance variation is conveniently measured by *standard deviation*, denoted by SD. Roughly, SD measures how far a *typical* occurrence of a random process will be from the average. For example, in many situations, about 68 percent of observations will be within one SD of the average. Often, the SD is difficult to compute, but there is simple formula for the SD of the number of *occurrences* of an event in repeated, independent trials, whether it be heads in coin tosses or "7" in dice rolls. If p is the chance of occurrence in one trial, then $1 - p$ is the chance of nonoccurrence, and the formula for the SD of, say, the number of occurrences in N trials is as follows:

$$SD = \sqrt{N \times p \times (1 - p)}$$

Since the chance that a coin lands heads is 1/2, to find the SD for the number of heads in, say, 100 coin tosses, we get:

$$SD = \sqrt{100 \times 1/2 \times 1/2} = 5$$

Since the average number of heads in 100 tosses equals $100 \times 1/2 = 50$, the observed number of heads in 100 tosses will *typically* be within one SD of the average, between 45 and 55.

When rolling a pair of dice, we have seen that the chance of rolling "7" is 1/6 and the chance of not rolling "7" is 5/6. Thus, for my computer simulation of 72,000 rolls,

$$SD = \sqrt{72,000 \times 1/6 \times 5/6} = 100$$

Since the average number of occurrences of "7" in 72,000 dice rolls is $72,000 \times 1/6 = 12,000$, a typical number of "7"s in 72,000 rolls would be within one SD of the average, or between 71,900 and 72,100 occurrences. In my 10 computer simulations of 72,000 dice rolls, 8 were in the one SD range.

Describing chance variation in units of SD from the average is made more precise by the central limit theorem.

The Central Limit Theorem

The central limit theorem, discovered by seventeenth-century French mathematicians, ranks with the law of averages as one of the great results in probability theory. It provides easy-to-compute, long-run probabilities for sums of repeated, independent trials of an experiment, in terms of the average and SD. A simple summary of probabilities provided by the central limit theorem can be stated as follows:

The number of occurrences of an event in a large number of independent trials of a process has about a 68 percent chance of being within one SD of the average, a 95 percent chance of being within two SD's of the average, and a 99.7 percent chance of being within three SD's of the average.

We'll apply this result to coin tosses. In 100 tosses, the average number of heads is 50 and the SD is 5. The central limit theorem says that there is a 68 percent chance that the actual number of heads will be between 45 and 55 (within one SD of the average) .There is a 95 percent chance that the actual number of heads will be between 40 and 60 (within two SD's of the average) and a 99.7 percent chance that the number of heads will be between 35 and 65 (within three SD's of the average).

In 72,000 dice rolls, the average number of "7"s is 12,000, and the SD is 100. Thus, there is a 95 percent chance that the number of "7"s will be between 11,800 and 12,200, and there is a 99.7 percent chance that the number of "7"s will be between 11,700 and 12,300. In my 10 computer simulations of 72,000 dice rolls, 9 were in the two SD range and all 10 were within the three SD range.

Statistical Logic and the Great Mishugi

Here is the key question in statistical decision making: *When is a weird event due to chance, and when is it due to something else?*

Answer: If an observed event is highly unlikely to occur due to chance and more likely under a plausible alternative, reject chance in favor of the alternative.

The Great Mishugi, a self-proclaimed psychic, claims to have mind-reading abilities. To test Mishugi, you toss a coin 100 times. After each toss, you secretly peek at the result, and Mishugi tries to guess it by reading your mind. When the experiment is complete, Mishugi has made 67 correct guesses. Does Mishugi have psychic powers, or are his results due to chance?

Using the central limit theorem, we have seen that in 100 coin tosses, there a 99.7 percent chance that the number of heads will be between 35 and 65. Thus, the chance is less than .003 that someone who is just guessing (with no psychic powers) will get a result as extreme as Mishugi's (more precisely, the chance of getting 67 or more correct guesses just by chance is about 1 in 2,500). We conclude that Mishugi's results are not just due to chance.

It turns out that Mishugi has been traveling around the country doing the coin-toss experiment with every scientist he encounters. By the lottery principle, even if Mishugi has no psychic powers, he will occasionally get weird results—results that seem to be outside the bounds of chance—just due to chance, simply because he's doing the experiment so often. It's tempting to overlook Mishugi's many poor showings and listen to him brag about his occasional successes, even if they only mean that Mishugi got lucky.

This brings up an interesting issue: Although Mishugi's activities in other towns would seem to have no effect whatever on the results of his coin-tossing experiment with you, the mere fact that he has been repeatedly doing this experiment should make you more skeptical of his current results. After all, he has to get lucky sometime. If you had more information about Mishugi's other attempts to prove that he is a psychic, you could be able to use that information in your current analysis.

In addition to Mishugi, there are dozens of other self-proclaimed psychics roaming around the countryside doing the coin-toss experiment. Even though these psychics have no interaction with either you or Mishugi, their existence clouds the waters of chance by providing more opportunities for success. Although their efforts don't make it any more likely that Mishugi will get lucky, you know that the more people who do the experiment, the more likely it is that *someone* will get lucky. Sometimes the world seems like a crowded casino, where occasional chance successes are highlighted by bells and whistles that make them stand out in a sea of failures. In this case, however, you are interested in Mishugi in particular, not just someone. He did better in his coin-tossing test than one would normally attribute to chance. Statistical logic compels you to find an alternative hypothesis under which Mishugi's results are more likely than chance allows. Psychic powers are one such alternative.

You decide to do a little more investigating and discover that Mishugi has done remarkably well whenever he does the coin-tossing experiment. You invite him to do the experiment again and secretly videotape the proceedings. Mishugi gets 71 correct in the second experiment, however, upon close inspection of the videotape, you discover that he has taped a small mirror to the ceiling that allows him to get a glimpse of the results of your coin tosses. When you conduct a third experiment by tossing the coin in a separate room so that Mishugi's cannot use the mirror, he makes 49 correct guesses. The alternative to chance that made Mishugi's triumphs less unlikely was his use of a mirror, not psychic powers.

Possible Job Discrimination Is Compared to a Box Model

A class-action lawsuit is filed against the police department of Anytown, USA. The plaintiffs claim that the department discriminates against female applicants. In the past five years, 40 percent of the qualified applicants for 100 new police-officer positions have been women, but only 20 of the 100 new officers are women. Police Chief Lamar says that the highest-qualified people were hired and that any discrepancies are due to chance.

To represent the fact that 40 percent of the qualified applicants were women and that 100 positions were filled from this labor pool, the statistical expert working for the women uses the following box model. Suppose a box contains 4 balls marked W and 6 balls marked M. One hundred balls are randomly selected from the box as follows: Each time a ball is selected, its gender (W or M) is noted and the ball is replaced in the box. What is the probability that 20 or fewer W's are selected?

The probability of interest can be found using the central limit theorem, in much the same way that we used it for coin tossing. Here, however, the chance that a W is selected equals .4, instead of .5 for heads in a coin toss. If you select 100 balls at random from a box containing 4 balls marked W and 6 balls marked M, the average number of W's equals $100 \times .4 = 40$, and the SD equals 4.90. (Don't forget, you put the selected balls back in the box before drawing new ones.) By the central limit theorem, there is a 99.7 percent chance of getting within three SD's of the average, which in this case is the range between 25.3 and 54.7 W's. Thus, the chance of getting 20 or fewer W's is less than .003. (It's actually about 1 in 50,000.) Statistical logic indicates that since this probability is so low, something other than chance is having an effect. Is it discrimination?

"Now, hold on just a minute," says Chief Lamar, who proceeds to point out that although the low probability may be true for the box model, job applicants aren't balls in a box. "In fact," claims the chief, "every person is a unique individual, not a ball in a box. The most qualified people were selected. The box model doesn't cut the mustard." The statistical expert agrees that the box model is only a model but notes that the model represents qualified applicants only. Chief Lamar agrees that 40 percent of the qualified applicants are women but says that some applicants are more qualified than others. At this point, Chief Lamar and the attorneys work out a deal in which the police department will hire more women, thus avoiding a costly court confrontation.

The box model in this example is intended to compare the police department's hiring pattern with what would happen if new officers were chosen completely at random from the pool of qualified applicants. I'm not suggesting that the department should select new officers at random from among the qualified applicants. The model serves as a baseline, or point of comparison, and can be used along with personnel records and other data to help the court determine whether or not the department is involved in illegal hiring practices.

The Monte Carlo Method

It's easy to compute probabilities for a coin toss or a dice roll, but some processes are too complicated to analyze with direct computations. In such situations, it is useful to construct a model of the actual process and then observe the results of a large number of identical repetitions of the model. The law of averages asserts that a large number of independent repetitions will provide accurate probability estimates. In this case the accurate probabilities are for the model, so their usefulness depends on how well the model simulates the real process. This procedure is called the Monte Carlo method, in honor of the gambling casinos of Monte Carlo.

The Monte Carlo method was first used by nineteenth-century mathematicians who manually repeated experiments a large number of times to obtain accurate probability estimates. This is tedious, and so the Monte Carlo method didn't become popular until the advent of computers. In the late 1940s mathematicians John von Neumann and Stanislaw Ulam used a computerized version of the Monte Carlo method in the development of the hydrogen bomb. Since then, the method has become an important tool for statisticians, economists, game theorists, physicists, and others who study complicated processes. In the 1950s statisticians used Monte Carlo techniques to develop winning strategies for the casino game of blackjack. These results shook up the gambling world with the publication of Professor Edward O. Thorp's book *Beat the Dealer.*

In my dice-rolling simulation, I used a computer to illustrate the law of averages by comparing simulated results with probabilities that were easy to compute. The Monte Carlo method is typically used to provide probability estimates in the absence of such computations.

Subjective Probabilities

Many processes aren't repeatable independently under identical conditions, and so the law of averages doesn't directly apply. The Monte Carlo method provides a clever way to add repeatability to the picture but is only useful if the model used for Monte Carlo simulation accurately depicts the actual process. In some situations, shrewd data analysis provides accurate probabilities even when the process doesn't satisfy the conditions of the law of averages. Such estimates are sometimes called *subjective probabilities.* The danger is clear: Without a mathematical structure, anybody can say anything. Let's look at some examples.

A stockbroker says there is a 70 percent chance that IBM will go up at least 10 points in the next month. The broker's probability is based on a careful study of market data. Although current market trends may be similar to past situations, the ups and downs of IBM are not repeatable under identical conditions like the heads and tails of coin tossing, and the law of averages doesn't directly apply. This doesn't mean that the stockbroker's opinion about IBM is wrong. If the broker's data includes all factors that significantly affect IBM, and if these factors can be modeled as a repeatable process, then even though some conditions change, the law of averages works.

A gambler studying football data has discovered that during the 1996 through 1998 football seasons, the Denver Broncos won (and "covered the point spread") in 90.9 percent of their games played on Sunday during the first 4 weeks of the season (10–1 record). Should the gambler bet on the Broncos on Sunday games during the first 4 weeks of future seasons? In other words, is the probability 90.9 percent that the Broncos will win such games in the future? The trend is true for 1996 through 1998, but the continually changing conditions of football create situations in which this pattern may not hold. What happens next year may be different from what happened in the previous 3 years. One way to get an idea of how the gambler's subjective probability will apply to future games is to test the trend on a different batch of historical data.

Chapter 3 🎲

The Many Faces of Chance

"God does not play dice with the universe."
ALBERT EINSTEIN

*A*ll creatures have strategies for coping with the uncertainties of life: Don't walk in front of a bus. Find a reliable water hole. Dig a burrow where predators won't find it. Don't get too close to that lion. Eat plenty of vegetables. Don't jump out of the nest until you can fly. Effective strategies enable a species to survive. As a natural extension of these strategies, humans have learned to be scientists by conducting experiments, developing theories, and making predictions based on measurements and mathematical models.

Chance Meets Classical Physics

While French mathematicians were developing the laws of chance, Sir Isaac Newton (1642–1727) was developing the laws of motion and universal gravitation, establishing the foundations of classical physics. For the next two hundred years, physicists viewed the world as a vast, deterministic system, a giant machine. According to this paradigm, to understand how a system works you need only to uncover the mathematical equations that govern it. Once these equations are known, you plug in appropriate numbers and determine, with certainty, the state of the system at any time in the past, present, or future. This was the scientific version of fate.

According to Newton's laws of motion, if you travel down the interstate at a steady speed of 65 miles per hour for 2 hours, you will cover a distance of $65 \times 2 = 130$ miles. The occurrence of a chance event, like getting stopped by the police or being struck by an asteroid, is treated as an event outside the process itself. When a physical process is fully understood, there is no randomness, or so the story went.

Newton's laws were amazingly successful and worked in a variety of situations, bolstering the deterministic credo. Failure to accurately describe a phenomenon was attributed to its complexity or to a faulty understanding of its underlying structure. Chance sometimes found its way into the picture, not as part of the official structure of a physical process but as a stopgap, to provide a temporary explanation of a complicated situation until greater knowledge could be obtained. In practical terms, this meant using the laws of chance to describe how objects behaved on the average, rather than knowing, with certainty, the behavior of every object in a system. For example, to develop laws describing thermodynamics and the motion of gases, nineteenth-century physicists made the assumption that molecules moved randomly. This was called Brownian motion, after biologist Robert Brown, who noticed the phenomenon in 1827. This assumption allowed development of useful mathematical models based on randomness that provided facts about the average motion of many molecules, rather than unknown, deterministic facts about the motions of individual molecules.

By using such statistical models scientists gained insights into complicated physical processes, but that didn't mean that chance was necessary. Classical physics still assumed that a natural phenomenon could be understood with certainty, no randomness, if enough information were known about its governing behavior. Taken individually, for example, molecules were still believed to move and interact in accordance with Newton's laws. There were just too many to monitor. The laws of chance were merely a mathematical convenience that approximated reality. It wasn't necessary to conclude that God rolled dice but only that rolling dice approximated what physics was currently unable to decipher. When technology advanced sufficiently, chance would be unnecessary.

As technology advanced, electron microscopes, X rays, and lasers allowed the observation and measurement of objects that were once invisible. With these spectacular inventions, it was reasonable to assume that, given the right equipment, anything could be measured precisely. This further supported the notion that statistical models were an artifice. Then, in the first part of the twentieth century, scientists made a series of startling discoveries that could be loosely summarized as follows: Chance, in some form, is a fundamental part of the way we perceive reality and can never be eliminated.

In the early 1900s, Neils Bohr, Werner Heisenberg, and other physicists studying atomic particles developed the laws of *quantum mechanics*. According to Heisenberg's famous *uncertainty principle*, it is impossible to simultaneously measure precisely both the position and momentum of a subatomic object. The more *precisely* you measure one thing (either position or momentum), the more *uncertain* you become about the other. This is not because of the crudeness of your measuring device, but it will happen with *any* measurement and, in fact, is caused by the observation procedure itself. In other words, on a microscopic level (and perhaps on any level), it is impossible for the observer to be completely detached from the observation. On a subatomic level, there's no such thing as the proverbial fly on the wall, unobtrusively watching how things unfold. The very existence of the fly, in this case, the scientist and measuring device, change the picture. It turns out that if you want to know the precise path of a subatomic object, the best you can do is give *probabilities* that the object will be in various locations. Quantum mechanics thus introduces chance into our basic perception of reality.

As the theory of quantum mechanics became experimentally verified, chance became an integral part of physics, formerly the citadel of certainty. Contrary to classical physics, the subatomic world is not like Sundays in the park, but instead it is a turbulent sea of randomness, where nothing stays still. In fact, the perceived *emptiness of space* is really just an average of quantum mechanical activity, in which energy fluctuations continually create and destroy matter, with positive and negative particles canceling themselves out almost as soon as they arise, averaging to zero. As Brian Greene puts it in his book, *The Elegant Universe*:

> If an energy fluctuation is big enough it can momentarily cause, for instance, an electron and its antimatter companion the positron to erupt into existence, even if the region was initially empty! Since this energy must be quickly repaid, these particles will annihilate one another after an instant, relinquishing the energy borrowed in their creations. . . . Quantum mechanical uncertainty tells us the universe is a teeming, chaotic, frenzied arena on microscopic scales. . . . Since the borrowing and repaying on average cancel each other out, an empty region of space looks calm and placid when examined with all but microscopic precision. The uncertainty principle, however, reveals that macroscopic averaging obscures a wealth of microscopic activity.

Over the years, skeptical scientists have unsuccessfully tried to prove that the chance in quantum mechanics is not really chance at all but uncertainty caused by unknown variables, complicated conditions, and mistaken assumptions. It is now generally agreed that chance is not only an integral part of reality but that the laws of chance must play a role in describing the nature of subatomic objects.

Since all things, from armadillos to zebras, are made up of subatomic matter, chance must be part of everything. So why shouldn't large objects have the same uncertainty properties as subatomic ones? Why do Newton's laws of classical physics work, for the most part, on a large scale? Why is it that we can accurately measure both the position and the momentum of a coffee cup but not of an electron? Aren't coffee cups made up of electrons and other subatomic objects?

There is a statistical explanation. In addition to the binding forces that hold objects together, coffee cups and other large objects are composed of so many subatomic particles that the chance is essentially zero that one will, say, spontaneously break apart into a cloud of molecular dust. It's like tossing a coin 100 times and getting heads on every toss: It may be possible, but it won't happen. The behavior of large objects may not be predictable with certainty, but the chances are so low that something really weird will happen that classical physics works. In other words, when enough electrons go walking, Newton does the talking.

Nonrandom Randomness: Chaos in Weather Prediction

"It was an extra-ordinarily bitter day, I remember, zero by the thermometer. But considering it was Christmas Eve there was nothing extra-ordinary about that . . . Seasonable weather for once, in a way . . .

It was a glorious bright day, I remember, fifty by the heliometer, but already the sun was sinking down into the . . . down among the dead . . .

It was a howling wild day, I remember, a hundred by the anenometer. The wind was tearing up the dead pines and sweeping them away . . .

It was an exceedingly dry day, I remember, zero by the hygrometer. Ideal weather, for my lumbago."

Samuel Beckett, *Endgame*

In 400 B.C. Aristotle wrote *Meteorological,* a book about the weather, from which we get the word *meteorology.* In the 1950s the prolific mathematician John von Neumann began to construct an enormous computer network to handle the many equations he believed to be necessary for accurate weather prediction.

Modern researchers like von Neumann didn't think that chance was an impediment to weather forecasting. Instead, they believed that inaccuracies in forecasting were due to two things: incomplete knowledge of the equations that determined the weather and insufficient computing power. Once these hurdles were overcome, accurate weather prediction would be as easy as predicting tides and the movement of the planets, long-standing scientific success stories. In 1961 an accidental discovery by a mathematical meteorologist ruined these ambitious plans.

In classic, Newtonian physics, there are two things necessary for the accurate prediction of a physical process: the correct equations for describing the process and good measurements.

You take the measurements, plug them into the equations, and get your accurate prediction. The effectiveness of this procedure sometimes depends on an assumption concerning the continuity of approximation: *If two sets of initial measurements are approximately the same, then predictions based on the measurements will also be approximately the same.* The continuity of approximation assumption is crucial, since all physical measurements are approximations. As we saw with the random trout, there's no such thing as an exact physical measurement.

Researchers like von Neumann believed that the key to accurate weather prediction was to develop more powerful computers and to build worldwide grids to monitor factors like temperature and barometric pressure with greater accuracy. This was done with some success, but accurate long-term forecasting of local weather conditions remained a challenge. Then along came Edward Lorenz.

In 1961 Lorenz, a research meteorologist at the Massachusetts Institute of Technology, used a clunky, Royal McBee computer to simulate weather systems. Using a simple set of equations to run his weather model, Lorenz input initial values and let his computer take the system through a series of iterations representing the development of weather patterns. As the system progressed, Lorenz's computer printed a graph of the weather.

One day, when Lorenz wanted to study a particular sequence of weather conditions, he plugged in the same initial conditions that he had used in an earlier experiment, expecting the new graph to fit the old graph and continue from where the old graph ended. What happened was quite different. At first the new graph fit the old graph, but then it dramatically diverged. This, thought Lorenz, was impossible. Both graphs started at the same place. Both were generated with the same initial values from the same set of equations. Therefore, they had to be the same.

Lorenz concluded that there was something wrong with his computer, but he soon realized that this was not the case. Finally, he confirmed what had happened: The starting measurements, or initial conditions, for the second experiment weren't exactly the same as for the first experiment. In the first experiment the starting numbers were carried out to six decimal places, while, for expediency, Lorenz had rounded off the second set to three decimal places. This shouldn't have mattered. This was a simple, deterministic system governed by a simple set of equations. There should have been continuity of approximation: Approximately the same initial conditions should have produced approximately the same future results, in this case, graphs. With Lorenz's equations, this didn't happen. This bizarre phenomenon dealt a blow to the widely held belief in classical physics that continuity of approximation was always true for physical systems.

Lorenz's weather equations violated continuity of approximation because they were sensitive to initial conditions: Even if you started with two sets of values that were approximately the same, as the process unfolded, the results became completely different. Mathematicians might dream up pathological equations that behaved this way, but here was a simple system developed by a meteorologist to model the weather. You would think that with slightly different starting conditions, say, slightly different temperatures or slight differences in the wind speed, a simple weather model like Lorenz's would develop in a way that was only slightly different. Instead, slightly different initial conditions meant totally different future weather patterns. In other words, Lorenz's simple weather model was inherently unpredictable: Approximate measurements of the real initial conditions could produce long-term predictions that were no more accurate than if a coin was tossed to predict the weather. Although Lorenz's weather model wasn't the real weather, it embodied important characteristics of the real weather and was presumably much simpler. Later experiments confirmed that Lorenz's model accurately portrayed many important aspects of real weather systems.

Processes that are sensitive to initial conditions are called chaotic. The impact of this characteristic in predicting the weather is sometimes called the butterfly effect: A butterfly flapping its wings in Paris today could affect the weather in San Francisco next month.

This is not just a mathematical curiosity. For a chaotic process in nature, even if there is a deterministic, mathematical understanding of how the process works, it is futile to make predictions without exact initial measurements. But that's impossible, since all we can do to measure natural systems like the weather is obtain approximations. We may improve the precision of our measuring devices, taking measurements to more and more decimal places, but we will never achieve perfect precision. Although short-term predictions may be accurate, if we try to predict far enough into the future, we are out of luck. This is a startling situation, because here we have a process that both obeys the laws of classical physics and also appears random, seemingly contradictory dynamics. Chaotic systems are Newton's worst nightmare: They are both deterministic and unpredictable.

After discovering the chaotic nature of his weather model, Lorenz wanted to see more. He used a simple set of equations to describe a water wheel. The wheel was made of buckets with holes in the bottom. Water poured in from the top. The rotation of the wheel varied according to the amount of water driving it. If only a small amount of water flowed into the top buckets, they would empty and the wheel would remain motionless. Nothing chaotic about that. If the flow was increased, the buckets filled enough to move the wheel. As the water pressure got stronger, the wheel moved faster. At some point the system became unstable. The buckets would move so fast that they didn't have time to fill at the top or empty when they started climbing the other side. As a result, the wheel slowed down, stopped, and changed directions entirely. If the force of water was strong enough, the process appeared random, with the wheel changing directions repeatedly and unpredictably, never in a predictable pattern. In this unstable range, the water wheel was chaotic, just like the weather model. Even though the equations describing the water wheel were simple, deterministic formulas, their sensitivity to initial conditions made long-term results appear random.

Chaos in Animal Population Growth

In the years since Lorenz discovered chaos in his weather model, chaotic behavior has been found in mathematical models of many processes, from biology to the stock market. For example, biologists have long been interested in understanding and predicting the growth of animal populations. In a simple mathematical model, called the logistic difference equation, the size of next year's animal population in a particular region is a function of this year's population size multiplied by a number that takes various other population factors into account. The mathematical properties of this equation were studied during the 1970s by biologist Robert May and mathematician James Yorke.

Expressing population size, popsize, as the fraction of total capacity of a region, the logistic difference equation can be written as follows:

popsize (next year) = R × popsize (now) × [1 − popsize (now)]

where

popsize (next year) = next year's population capacity
popsize (now) = this year's population capacity
R = a number that can vary between 0 and 4.

For example, popsize (now) = .7 indicates that this year's population is 70 percent of capacity. The term [1 − popsize (now)] keeps the model from displaying unlimited growth, a natural population dynamic that is affected by factors like available food and space. If R is greater than 4 or less than 0, the resulting process can go outside the 0–1 measurement scale used to represent population capacity as a fraction.

As in Lorenz's weather and water-wheel models, the logistic difference equation is *nonlinear,* because its graph isn't a straight line. Processes governed by nonlinear equations, even simple ones like the logistic equation, can exhibit chaotic behavior.

For values of R less than 3, the logistic equation behaves nicely. In this case, a population can vary in size from year to year, but it quickly reaches an equilibrium, or stable state, known as an attractor. For example, when R is between 0 and 1, every starting population will become extinct, eventually getting arbitrarily close to 0.

If R = 2, every starting population will eventually reach the value .5 (50 percent of capacity) and stay there. For example, suppose R = 2 and the starting population equals .7. Then, by plugging values into the logistic difference equation, we see the following population growth:

year 1: popsize = .7

year 2: popsize = 2 × .7 × .3 = .42

year 3: popsize = 2 × .42 × .58 = .4872

year 4: popsize = .49967232

year 5: popsize = .49999979

year 6: popsize = .50000000

and .5 forever after

If R = 2.5, every population will eventually reach the stable state of .6. If R is between 3 and 3.5, the starting population still will eventually stabilize, but now there is more than one equilibrium state. For example, if R = 3.1, every starting state will eventually cycle through two equilibrium points, .7646 and .5580. If R = 3.5, the population will eventually cycle through four states, .8750, .3828, .8269, and .5009. This is the beginning of instability.

When R is greater than 3.5, the process goes haywire. The number of attractors starts doubling, from 4 to 8 to 16 to 32 to 64, and so on, until, for values of R close to 4, the population size never settles down but moves instead through an infinite range of values. When R = 4, the process appears random, traveling in an unpredictable pattern among the entire possible range of population values from 0 to 1. Like Lorenz's weather model, the most accurate starting measurements are only approximations of true population capacity and are thus useless for predicting long-run results. A change in the tenth decimal point of current population size can make a big difference in the population's long-term behavior. A frog's unlucky demise today can theoretically make the difference between a pond full of frogs and no frogs at all fifty years from now. In other words, when R = 4, any population whose growth is determined by the logistic equation may as well be governed by tossing coins to decide whether the population eventually survives.

There are many other dynamics, such as pollution and human encroachment, that affect the growth of animal populations. Even populations controlled by the logistic equation only show odd behavior when the factor R is close to 4. Thus, population growth that follows the logistic model can be a predictable process, or it can exhibit weird, unpredictable behavior when parameter values are in chaotic zones.

Chance Devices Are Chaotic

Tossing coins, rolling dice, and other simple processes that are used to demonstrate randomness are themselves chaotic processes. For example, if you can measure the exact angle that dice hit a table, the exact velocity with which you roll them, and other important initial conditions, it is theoretically possible to predict the outcome of the roll. Unfortunately, dice rolls are chaotic, and sensitive enough to initial conditions that gamblers are unable to make accurate predictions.

Magicians, charlatans, and even some statisticians (the categories aren't mutually exclusive) can partially control the results of coin tosses. The precise location of the coin in your hand, the place where your thumb hits the coin, and the velocity of the spin are initial conditions to which the coin toss is sensitive, making coin tossing another chaotic process.

Professor Edward O. Thorp, author of the classic blackjack book *Beat the Dealer*, once teamed up with renowned mathematician Claude Shannon to develop electronic equipment that could predict the results of roulette spins. They were able to track the trajectory of the ball after it was tossed into the spinning wheel and predict with some accuracy in which part of the wheel the ball would land. This technique, described in Thorp's book *The Mathematics of Gambling*, takes some of the chance out of chance and demonstrates how a roulette wheel, a supposedly chance device, may also be a chaotic system.

Many of the devices we mortals use to generate randomness are chaotic systems that can theoretically be described by deterministic, mathematical equations. These devices generate a version of chance that works fine for our purposes, but this might be different from the pure chance that manifests itself in the subatomic world. Maybe Einstein was right: God doesn't roll dice. Humans roll dice. It's the best we can hope for when we try to imitate pure randomness.

Chaos Can Be Stable

An interesting feature of chaotic processes is that there are ranges of stability. For example, the logistic equation for predicting animal populations is only unpredictable for values of R close to 4. Otherwise, the population stabilizes to a limiting state or states. If dice are rolled very slowly, without any spin, it is easy to determine how they will land. The same is true for a slow-moving roulette wheel, for coins that don't spin, for low levels of water pressure in Lorenz's water wheel, and for certain parameter values in the weather model. It's only when systems get turbulent that unpredictability sets in.

Models and Metaphors

Although a rose is a rose is a rose, and dice are dice are dice, Lorenz's weather model is not really the weather and the logistic equation is not really animal population growth. Instead, they are mathematical models of natural processes that we hope will provide us with an understanding of how these complex processes work. In our never-ending quest to understand reality, we are often limited to models and metaphors.

For centuries scientists believed that natural processes were inherently predictable, even though occasionally things didn't work quite right. It is now apparent that the occasions when things didn't work quite right were not just weird anomalies but fundamental aspects of the situation. You can completely describe a system. No randomness is involved; everything is cause and effect. But because of sensitivity to initial conditions and the fact that all physical measurements are only approximations of unknowable, exact values, the system is unpredictable. Maybe you can accurately predict a sunny day in the desert or snow at the North Pole, but whether it will rain in Miami three weeks from next Tuesday is anybody's guess.

Initial Conditions and Daily Life

If a process is sensitive to initial conditions, then it is sensitive to all conditions, since every condition is an initial condition for the process starting at that point. If, in some general sense, life is a chaotic process, then seemingly insignificant events, minor changes to your daily routine, little things that you can't control can dramatically alter the course of your future. In a world driven by chaotic forces, even if you completely understand the physics of these forces, some important aspects of your life may be as unpredictable as if nature tossed a coin to decide your fate.

Chapter 4 🎲

Using Chance
to Get Information

"She loves me, she loves me not . . ."
ANONYMOUS, LOVESICK PETAL PICKER

*L*ong before modern mathematicians developed probability theory, ancient sages used chance devices to consult the gods for advice. Rolling dice to get a divine sign may appear silly compared with modern, sophisticated information-gathering techniques, but these rituals were the precursors to random sampling, an important tool of modern scientists.

It wouldn't have made much sense for ancient sages to attempt to compute the probability that messages they received from the gods were correct. The gods were *always* correct, and the sages were the liaisons between the gods and the people. Thus, if the sages wanted to keep their jobs, it was important that the advice they obtained from the gods was phrased so that it was, in fact, always correct. This resulted in metaphorical messages and carefully crafted oracles that were so general they would always be true. This has its value. Not all wisdom is meant to predict whether IBM will go up ten points next month.

The *I Ching:* Tossing Coins to Predict the Future

"We must admit that there is something to be said for the
immense importance of chance."

> Carl Gustav Jung, Foreword to the *I Ching*
> (Wilhelm and Baynes version)

One of the earliest known systems of using chance to consult
the gods for advice is the *I Ching,* or *Book of Changes,* an
ancient book of divination developed by Chinese sages thousands of years ago and still in use today. The *I Ching* is a
sophisticated treatise whose procedures, subtleties, and layers of meaning make pop philosophy seem banal. In order to
select an *I Ching* passage, the spiritual seeker formulates a
question and performs a ritual to select a passage that a crass
Westerner might liken to a lottery: the mystic quick pick. On
the other hand, someone in tune with Eastern mysticism
would assert that chance provides a way to tap into the flow
of the universe.

The great Swiss psychologist Carl Jung, who had a
healthy respect for the spiritual side of chance, asserted that
the *I Ching* was based on the principle of synchronicity (his
term), which asserts that all things in the universe, even seemingly unrelated events, are somehow connected. As Jung put
it, "synchronicity takes the coincidence of events in space and
time as meaning something more than mere chance, namely, a
peculiar interdependence of objective events among themselves as well as with the subjective states of the observer or
observers." The notion that objective events are intercon-
nected with the subjective states of the observer is now
accepted by modern science and, in particular, is an important part of Einstein's theory of relativity, not to mention
quantum mechanics. As for trying to tell if seemingly unrelated events are connected, we have seen that chaotic processes like the weather are so sensitive to initial conditions
that seemingly unrelated events are actually related and can
have an important impact on the future. On the flip side of
the coin, given enough opportunity, models of pure random-

ness will yield patterns so bizarre that they seem to have been caused by an underlying mechanism other than chance.

To consult the *I Ching*, the seeker of wisdom silently asks for advice about some topic and then tosses coins, yarrow sticks, or tortoise shells or uses some other chance device to randomly pick one of 64 hexagrams. Each hexagram consists of six yin-yang lines that correspond to a particular passage of *I Ching* wisdom that will presumably shed light on the seeker's query. Chance is the mechanism with which the seeker gives up conscious choice, plugs into the flow of the universe, and gets a message from higher powers.

There are many English translations of the *I Ching*, for example, the scholarly version by Richard Wilhelm and Cary F. Baynes (Princeton University Press, 1950). I will use *The I Ching Made Easy*, a down-home version by Roderic Sorrell and Amy Max Sorrell (HarperCollins, 1994), which has a simpler selection procedure and less prose than the scholarly version.

Here is the question I asked the *I Ching*: "Should I take a break from being a statistics professor to manage the university television station?" Actually, I had been managing the television station for two years when I posed this question, but I thought it would be a good idea to get the oracle's opinion, even after the fact. I was initially worried that I would be cheating if I asked the *I Ching*'s opinion about something that had already happened, however, upon reflection, I decided that since the *I Ching* provides information about the totality of a situation, cheating is impossible.

I tossed my yin-yang coins, consisting of two quarters, three dimes, and a nickel, and appropriately arranged them to select a hexagram. The result was tails, tails, heads, tails, heads, heads, yielding Hexagram 54, which had the following title and description:

Hexagram 54
Impulsive, Passionate, Flawed
Fools rush in.
Advancing to war brings misfortune.
Thunder over the lake. The lake represents the desire to
break free of restraints and come forth.
The thunder represents a sudden, instinctive rush, as in
sexual excitement. Together they imply a very energetic
but possibly rash and premature display.
You are in danger of taking on too much too soon. Emotions
make you hasty and reckless. However, nothing ventured,
nothing gained.
Having the courage to take risks is what life is about.

Using the fact that my nickel landed in the fifth place,
I obtained the following additional admonition:

The bride is more modestly dressed than the bridesmaids.
You have no need to show off or be the big shot. When you
have the upper hand, stand back and let others shine.

Except for the part about the bridesmaids, the passage is
meaningful to me. Two years ago I pondered these same
issues. I was well aware of "the danger of taking on too much
too soon," and I knew from years of practice that "emotions
make you hasty and reckless." On the other hand, I believed
that "nothing ventured, nothing gained" and that "having
the courage to take risks is what life is all about." Anyway,
after weighing the pros and cons, clearly articulated two
years later by Hexagram 54, I made the move.

Since the I Ching is consulted by people in a myriad of situations, its passages must be open to many interpretations.
Jung suggests that the phenomenon of different people having
different interpretations of the same I Ching hexagram is not a
flaw but a necessary part of the process. Hexagram 54 seems
to address the main issues of my personal decision problem,
but it undoubtedly also addresses the main issues of many
other personal decision problems. Since the user of the I
Ching is part of the process (remember, there is no such thing
as an objective observer), it is logical for different users to

have different interpretations of the same passage. Like all good oracles, *I Ching* passages are worded with great metaphysical generality, so in some sense the *I Ching* can never be wrong. One might suggest that such generality makes the *I Ching* meaningless, however, any information that gives the seeker new insights could hardly be called meaningless. Even two years after the fact, the *I Ching* helped me clarify my earlier decision. As the old saying goes, "Don't pooh-pooh the oracle just because it doesn't predict the winner of the fifth race at Santa Anita."

A few years ago, in a bizarre blend of the old and the new, a Southern California video-game company developed a hand-held computer game called the *I Ching Lottery Predictor* that randomly selected lottery numbers using a computerized version of tossing yin-yang coins. At the time of this writing, the company is in bankruptcy proceedings. Such is fate.

The World Is an Oracle
Random Sampling

Although the *I Ching* is still widely used throughout the world, tossing coins to consult the gods is not considered an S.C. (statistically correct) way to get information. This is not because of the use of chance to find the appropriate *I Ching* passage but because the *I Ching* may not provide the type of information model that today's seeker of knowledge desires. A different approach is to toss coins to select a representative sample from a population under study. It doesn't matter whether the population consists of people, pandas, or parts coming off a production process. This technique, known as random sampling, requires the use of a chance device, like a coin or a computer's random-number generator, to insure that everyone or everything in the population under study has the same chance of being selected. The simple technique of random sampling is often necessary due to the immense size or inaccessibility of the population being studied. It provides an amazingly accurate method of obtaining information about an entire population from a small sample. For example, the Gallup Poll, CBS-*New York Times* Poll, and other polling organizations make consistently good estimates of national public opinion based on samples of about 1,000 randomly selected people. Another application of random sampling is quality control, in which the objects sampled are products, not people. By taking a random sample of the items in a production process, quality-control engineers can quickly determine when there is something wrong with the process.

Even though modern random-sampling techniques are performed for practical purposes, those who believe in the use of chance to get information should be as respectful of the *I Ching* as they are of the Gallup Poll.

Nonrandom Sampling

Never underestimate the importance of randomness when trying to obtain a representative sample. When a nonrandom sample is used to obtain an estimate, the results can give a grossly distorted picture. Newspapers and Web sites often conduct polls in which the reader or viewer is asked to voice an opinion by phone or Internet. Such surveys are based on self-selection rather than random selection: Whoever feels like participating may do so. In order to demonstrate how easy it is to distort a nonrandom sample, I participated in an Internet poll in which viewers at a certain Web site were asked to vote on whether NATO should stop bombing Yugoslavia. In five minutes of feverish electronic voting, I single-handedly raised the percentage who wanted NATO to stop bombing from 53 percent to 59 percent. Even when nobody is cheating, because the respondents to a self-selection survey may have a strong opinion and may not be representative of the entire population, such surveys are often inaccurate and shouldn't be trusted.

Random-Search Software

In Chapter 2 I described a gambler who discovered that during the 1996–1998 pro football seasons the Denver Broncos won 90.9 percent of games played on Sunday during the first four weeks of the season (10–1 record). Time will tell whether this unusual win-loss record is a random fluke or an accurate predictor of certain games that will reap rewards for savvy sports bettors.

The pattern was obtained with software that searches randomly through computerized football data, automatically finding weird patterns. In this case, it is the patterns themselves that are randomly sampled. This is extremely tedious to do manually, even in a limited way. For example, to find how the Denver Broncos did on Sunday during the first four weeks of the season, you would have to tally the results of many seasons, somehow uncovering the interesting pattern that arose during the first four weeks.

In automatically finding patterns, the computer "thinks" of things that we would never think of ourselves, even if we had the time. The software doing this search discovered and analyzed almost a million patterns per hour.

Such awesome computing power can yield misleading results. In particular, you run smack into the lottery principle: When you look at enough patterns, you will find weird ones caused by chance alone that have no predictive power. For example, the chance of a team having a 10–1 record in certain situations, just due to chance, is about 1 in 200. With enough data and computing power, finding such an oddity caused by chance alone is like catching a hungry trout in a well-stocked pond. It becomes necessary, therefore, to see if your bizarre patterns still hold true against a fresh set of data before declaring them useful.

Focus Groups

Random sampling and random searching are not the only neat ways to use chance to get information. Another popular method of probing either data or the public consciousness is the focus group, in which the researcher gathers together a group of people to explore some topic of interest. Unlike a random sample, in which respondents are typically asked carefully prepared survey questions, a focus group calls for members to engage in freewheeling discussions. In this way, the chance comments of participants may provide the researcher with new insights that wouldn't have arisen in a more formal context. It's like randomly searching through people's minds about a specific topic.

If you want new insights on life, new understandings of issues that bother you, try running your own, informal focus groups. You can do this by talking to friends and acquaintances, asking their opinions about things. You'd be surprised at how many problems you can solve just by talking to people. You can also get inspirations and solve problems by randomly wandering around town, taking in new scenery. Let the world be your *I Ching*. Use chance to get information.

Chapter 5 🎲

Roulette:
A Civilized Way to Lose

"I hope I break even this week. I need the money."
VETERAN LAS VEGAS GAMBLER

*T*he perfect place to observe the laws of chance in action is a casino; in the carefully controlled confines of the citadels of chance, reverence is paid to randomness through rituals of risk known as gambling games. A few casino games allow the player to exercise a modicum of skill, but most are games of pure chance, in which the gambler must rely on luck to walk away a winner. In this chapter we'll analyze roulette, a leisurely game of pure chance and panache, where you can relax and savor the sultry air (or choke on the smoke) of cool casino decadence.

> "Feeling as though I were delirious with fever, I moved the whole pile of money to the red—and suddenly came to my senses! For the only time in the course of the whole evening, fear laid its icy finger on me and my arms and legs began to shake. With horror I saw and for an instant fully realized what it would mean to me to lose now! My whole life depended on that stake!"
>
> Fyodor Dostoyevski, *The Gambler*

Roulette Basics

Roulette is played with a horizontal wheel that spins on a shaft mounted next to a betting layout. In American casinos the wheel is divided into 38 sections, numbered from 1 to 36, 0, and 00. (In European casinos there is no 00.) Of the sections numbered from 1 to 36, 18 are red and 18 are black, with the colors alternating; the sections marked 0 and 00 are green and are on opposite sides of the wheel from each other. A casino employee spins the wheel and then tosses a ball along the wheel in the opposite direction of the spin. As the wheel and ball slow down, the ball lands in the winning section. A gambler bets by placing chips on the betting layout, usually before the ball is tossed into the wheel.

There are numerous roulette bets. Here are two:

- Bet that the winning section is red. The payoff odds for a red bet are 1 to 1, or even money. This means that if you bet $1 on red and win, you get your dollar back, plus a dollar profit. Since 18 of the 38 sections on the wheel are red, the probability that you'll win a red bet equals 18/38.

- Bet that the winning section is 17. For a single-number bet, the payoff odds are 35 to 1. This means that if you bet $1 on 17 and win, you get your bet back plus a $35 profit. Since there is only 1 section marked 17, the probability that you'll win a bet on 17 equals 1/38.

Which of these two bets is better? One answer is that it depends on what gets you excited. But let's quell the quick quips and be quantitative.

Payoff Odds

In order to determine the long-run results of a roulette bet or any other situation that involves repeated wagers under identical conditions, you must know two things: probabilities and payoff odds. The payoff odds for a particular proposition give the amount by which you profit, usually expressed per dollar bet, should you be so lucky as to win your wager. Although reasonably regular in respectable betting establishments, payoff prices are arbitrarily set by the casino and are usually slightly lower than the true odds. In this way the casino makes a long-run profit yet the gamblers aren't discouraged. Since there are 18 red sections and 20 nonred sections, the true odds against winning a red bet are 20 to 18, whereas the payoff odds are 1 to 1, slightly less than 20 to 18. In the case of betting on 17, there is one section marked 17 and 37 other sections: The true odds are 37 to 1, whereas the casino payoff odds are 35 to 1.

Bets on Red

The law of averages reveals that in repeated play, red will come up an average of 18 spins in 38 and not come up an average of 20 spins in 38. Since the payoff odds are 1 to 1, if you always bet $1 on red, you will win an average of $18 and lose an average of $20 per 38 spins, for an average loss of $2 per 38 spins, equivalently, 2/38 = 5.3 cents per dollar bet. This number is a weighted average, computed by weighing the chances of winning against the payoff odds. It is called the player's expected payoff. In this case, the player's expected payoff equals –5.3 cents per dollar bet, indicating that in repeated play, the persistent gambler who always bets on red will lose, at the rate of 5.3 cents per dollar bet. Dropping the minus sign and stating this number as a percentage, we get 5.3 percent, which is known as the house edge. The house edge is the casino's expected payoff, namely, its long-run average profit for this bet, expressed as a percentage.

The gambler's expected payoff or, equivalently, the casino's house edge measures the worth of a wager because it accurately predicts long-run average winnings and losses. To systematically compute a player's expected payoff for a given bet, you can make a table of possible payoffs with their probabilities and take a weighted average by multiplying the payoffs by probabilities and summing.

$1 Bet on Red

Win	Probability
1	18/38
−1	20/38

Expected payoff = $(1 \times 18/38) - (1 \times 20/38) =$
$-2/38 = -.053$

Bets on 17

What about bets on 17? The law of averages says that in repeated play, 17 will come up an average of 1 in every 38 spins and not come up an average of 37 spins in 38. Noting that the payoff odds are 35 to 1, if you repeatedly bet $1 on 17, you will win an average of once in 38 bets, for a profit of $35, and lose an average of 37 times, for a loss of $37. This yields an average loss of $2/38 = 5.3 cents per dollar bet. The player's expected payoff is −5.3 cents. House edge is 5.3 percent, the same as bets on red.

From the perspective of expected payoff, bets on red and bets on 17 are the same. In fact, all but one of the many possible roulette bets have a 5.3 percent house edge: In repeated play the persistent bettor will lose and the casino will make a profit of 5.3 cents per dollar bet.

From the casino's perspective, the long run doesn't mean that a particular player must play for a long time; it only means that there is plenty of action. One player making 100 bets is the same as 100 players making one bet each. Since, for example, it doesn't take long for one player to make 100 bets, the long run really isn't very long no matter how you look at it.

Roulette Strategies

Anyone who's watched roulette or any other casino game knows that, regardless of the law of averages, there is chance variation from play to play. Gamblers have winning streaks and losing streaks. Numbers run hot and cold. Perhaps the dire, long-run results predicted by the law of averages can be avoided by clever short-run play: You go in, you make a profit, you get out, nobody gets hurt. Here are some examples of roulette strategies.

Simultaneous Bets

Individual roulette bets give the casino a 5.3 percent edge, but what about making many bets at once? Instead of just betting on red or just betting on 17, how about betting on lots of things on the same spin, thus covering different possibilities and increasing your chances of winning?

Although making many bets increases your chances of winning, it also increases the total amount you bet. It turns out that averages are additive: Your expected payoff for many bets is the sum of your expected payoffs for each individual bet. If you lose an average of 5.3 cents per dollar bet on each individual bet, the same will be true for any collection of these bets. Moral: You can't turn a group of bad bets into a good one. If you try, you'll end up like the retailer who sold everything at a loss, hoping to make a profit from the increased volume.

Doubling Up

This popular strategy is used by gamblers and investors the world over. Here's the roulette version:

Bet $1 on red. If red comes up, you win, in which case you quit with a $1 profit. If red doesn't come up, double your first bet and put $2 on red. If red comes up, you win $2, in which case you should quit, thereby covering your previous loss and leaving you a $1 profit. If red doesn't come up, double your second bet and bet $4 on red. If red comes up, you win $4, in which case you should quit, thereby covering your previous losses of $2 and $1 and leaving you a $1 profit. If red doesn't come up, double your third bet and bet $8 on red. If red comes up, you win $8, in which case you should quit, thereby covering your previous losses of $4, $2, and $1 and leaving you a $1 profit. If red doesn't come up, make a fifth bet, and so on.

If you use the double-up strategy, red is certain to come up eventually, at which point you quit a winner. Right? Wrong. Casinos have both minimum and maximum bets on all games. The minimum bet is to keep desperate clowns from betting a penny. The maximum bet is to keep wealthy degenerates from betting enough money to bankrupt the casino if they get lucky. Why should a casino risk going broke when the law of averages guarantees that it will make a steady profit from many small bets? With the double-up strategy, if you have an unlucky losing streak, you may wind up having to bet more money than the betting maximum or your own maximum allows, just to cover your losses. For example, if red fails to come up fifteen times in a row, on the sixteenth bet you must wager $32,768 to come out $1 ahead.

Betting limits, both yours and the casino's, may seem like a minor bump on the road to success when using the double-up strategy. In fact, the minor bump ruins the strategy. Repeated plays of this system are like a shark in a school of goldfish: occasional large losses swallow many small profits.

Quit While You're Ahead

If you only use the double-up strategy once or twice and actually quit, per instructions, never to play again for the rest of your life, you will most likely win a dollar or two. This is contrary to the gambler's definition of quitting: Stop gambling for an hour while you have dinner at the buffet.

Unfortunately, if you keep playing the double-up strategy, in the long run you will lose at the same rate as any other persistent roulette bettor, 5.3 cents per dollar bet. In fact, doubling up is not a good strategy for any gambling game.

Bet on Numbers That Are Due

Suppose you watch the roulette wheel until red comes up ten times in a row and then bet on black, because black is *due*. Unfortunately, the law of averages doesn't say anything about what will happen on a particular spin, so a streak of reds does not mean black is any more likely to come up than it ever was. As the lottery principle predicts, weird streaks and bizarre patterns will occur by chance alone. If you see red come up ten times in a row, it's probably a random fluke. In a game like roulette, nothing is ever due.

Bet on Numbers That Are Hot: The Biased Wheel

Suppose you watch the roulette wheel until red comes up ten times in a row. This time, however, you bet on red, because red is *hot*. Again, weird streaks and bizarre patterns will happen by chance alone. If you see red come up ten times in a row, it's probably a random fluke; however, if there's something strange going on that causes a disproportionate number of reds to come up, it makes sense to bet on red, not black. Maybe an equipment malfunction has escaped the casino's notice. If the roulette wheel is warped, an opportunistic bettor can exploit the situation like a hyena who finds a wounded gnu.

I once heard a story about a couple of crazed gamblers who embarked on a quest through Nevada to find a warped roulette wheel. After more than a year of searching, they found one. After a few hours of heavy betting, the gamblers won thousands of dollars. Eventually, the pit boss got suspicious and shut down the wheel. The gamblers were happy. They had experienced a winning streak most gamblers only dream about. Their profit, taking into account the desperate months they spent searching for the warped wheel, averaged out to $3 an hour.

How can you tell if a wheel is warped? Let's say there is an extreme warp in section 17, so that the chance that 17 comes up is 2/38 instead of 1/38. To detect this just by watching the wheel, you would need to watch a large number of spins to see if 17 comes up more often than it should. In order to make such a decision you must account for chance variation. One way to do this is by computing the standard deviation and using the central limit theorem, which we discussed in Chapter 2. It turns out that even for this extremely warped wheel you need to observe about 2,000 spins to get enough separation to effectively rule out chance: In 2,000 spins, the three-SD range around the average number of 17's for the warped wheel (average = 105.26; 3SD = 29.96) does not overlap the three-SD range for an ordinary wheel (average = 52.63; 3SD = 21.48). In other words, the number of occurrences of 17 with this extremely warped wheel will most likely be outside the three-SD range of an ordinary wheel, so you could confidently rule out chance variation. Unfortunately, roulette is a slow-moving game. If there is one spin every 5 minutes, it will take about a week of continuous wheel watching to observe 2,000 spins.

There is yet another problem: If you are observing roulette games, looking for a warped wheel, you have no idea in advance that there is a dip in section 17, so you would have to keep track of all 38 sections. This opens the door for the pervasive lottery principle: Since you are now not looking for a particular warp, but only for some warp, there is more opportunity for weird results due to chance alone. Thus, you would have to either take more observations or use a more sophisticated statistical procedure in order to ascertain that a wheel had a warp.

Use a Computer

As mentioned in Chapter 3, when Edward O. Thorp teamed up with renowned mathematician Claude Shannon to develop equipment to beat roulette, they used a computer. They tracked the trajectory of the ball after it was tossed into the spinning wheel and predicted with some accuracy in which section of the wheel the ball would land. This procedure, described in Thorp's book *The Mathematics of Gambling*, took some of the chance out of the game. Unfortunately, the system required the use of sophisticated electronic equipment in the casino as well as placing bets after the ball was tossed into the spinning wheel. Casinos don't appreciate such behavior, especially if you're winning; however, computer technology now allows sufficient miniaturization to effectively hide your gear when employing this system.

Goals

In order to compare strategies for any game, the player must have a specific goal. This important aspect of game analysis is usually overlooked in the casino, where all gamblers are assumed to have the same goal: winning money. Different goals can yield different optimal strategies for the same game, however, and since games like roulette don't offer much hope for long-run success, it may be prudent to consider alternative goals.

Perhaps you are spending the weekend in Atlantic City. Realizing that you will lose in the long run, you just want to have fun for as long as you can before you go broke.

For games like roulette, in which the casino has an edge, "timid play" is the optimal strategy for maximizing average playing time: Always make the minimum allowable bet at the table of your choice. If the minimum bet is $1, always bet $1. If the minimum bet is $5, always bet $5. You will lose at the lowest rate possible, thus maximizing your playing time. This strategy is the casino version of the hourly-wage method of ranking jobs: Just as a job hunter chooses the job with highest hourly wage, the gambler picks the strategy with smallest hourly loss. For example, even though roulette bets have a worse house edge than some other gambling games, roulette is a leisurely activity, and the timid player may be "in action" (still betting) much longer than a gambler playing a faster-paced game.

On the other hand, to maximize the chance of reaching a specific monetary goal, "bold play" is optimal. Suppose you owe a loan shark $1,000. You must pay by 4:00 P.M. today or face dire consequences. You have $200 to your name. You need to reach your goal of $1,000. Nothing else matters. You decide to play roulette and bet on red. Your best chance of coming up with $1,000 is bold play: Always bet everything necessary to reach your goal. To start, bet your entire $200 on red. If you get lucky and win, you now have $400, which you then bet on red. If you win again, you have $800. Since your goal is $1,000, you should now bet only $200. If you lose, you now have $600, so you should bet $400. If you lose, you now have $200, and you again bet it all. If you lose, you are broke, and you better start looking for a good HMO.

The need to have a specific goal is particularly important in games outside the casino, where there may not be an obvious goal, like winning money. For example, a military strategy may be quite different for a country whose goal is to seize another country's natural resources than for a country whose goal is for the other country to establish a democracy.

Summary

Roulette is a game of pure chance (no skill, none whatsoever), with a house edge of 5.3 percent. It is mathematically impossible to win in the long run, no matter what strategy you use. On the other hand, you may be happy handing control of your money to the fates and getting free drinks as part of the deal, knowing that the only way you can walk away from the roulette table a winner is if you get lucky.

Chapter 6 🎲

Craps: A Relentless Ride through the Rapids of Risk

"You just have to wait until the dice get hot."
ANONYMOUS DEGENERATE GAMBLER

Craps is a fast-paced, high-energy dice game played on a classy table with a complex betting layout. Watching a crowded craps game, you see a seething scene of chance variation. The fortunes of the bettors oscillate wildly. Some players win, others lose; nothing ever settles down. If you observe the ups and downs of a single, persistent player, however, you will notice that eventually the law of averages ruthlessly imposes its will and that despite the turbulence of chance, the bettor's bankroll inexorably diminishes.

The Setup

Bets are placed on the outcomes of dice rolls. The game is typically run by four casino employees: the box man (who puts your money in a box), the stick man (who rakes in the dice with a stick), and two dealers (such a deal). There is also a pit boss, who roams around giving cold stares. Bettors stand around the table and place chips, which they've bought from the dealer or box man, on the betting layout.

One of the bettors, called the shooter, rolls a pair of dice. This responsibility rotates among the bettors at the table. Don't be shy when it's your turn. Shake the dice and fling them on the table, trying to hit the table wall opposite you. It's especially fun to aim your dice at piles of chips other bettors have placed on the betting layout. Don't deliberately throw the dice off the table, however, or you'll annoy the pit boss.

Unlike roulette bets, which give the casino a 5.3 percent edge, craps bets offer a variety of house edges, some worse than roulette, some better. In order to analyze craps bets, you need to know dice probabilities. Recall from Chapter 2 the 36 dice combinations (y = yellow die, b = blue die):

y,b	y,b	y,b	y,b	y,b	y,b
1,1	1,2	1,3	1,4	1,5	1,6
2,1	2,2	2,3	2,4	2,5	2,6
3,1	3,2	3,3	3,4	3,5	3,6
4,1	4,2	4,3	4,4	4,5	4,6
5,1	5,2	5,3	5,4	5,5	5,6
6,1	6,2	6,3	6,4	6,5	6,6

Most craps bets involve the sum of the upturned dots. It's easy to compute probabilities. Just use the table to add up the combinations for each sum (each set of combinations for a specific sum is on a diagonal of the table) and divide by 36, the total number of combinations:

Sum of Dots	Number of Combinations	Probability
2	1	1/36
3	2	2/36
4	3	3/36
5	4	4/36
6	5	5/36
7	6	6/36
8	5	5/36
9	4	4/36
10	3	3/36
11	2	2/36
12	1	1/36

A Sampler of Craps Bets

Betting on 7 and Other One-Roll Bets

Some craps bets are easy to describe. Betting on 7, for example, is a no-brainer: Place chips in the box marked SEVEN in the center of the betting layout. If the next roll of the dice is 7, you win. Otherwise, you lose. Payoff odds are 4 to 1.

Let's say that you repeatedly bet $1 on 7. The chance is 1/6 that 7 comes up and 5/6 that it doesn't. The law of averages asserts that you will win an average of 1 bet out of 6, for a profit of $4 (4 to 1 payoff odds) and lose an average of 5 bets out of 6, for a loss of $5. This yields an average loss of $1 for every 6 such bets, equivalently, a loss of $1/6 = 16.7 cents per dollar bet. In other words, the house edge for bets on 7 equals 16.7 percent, giving a huge advantage to the casino, about three times bigger than roulette. You might as well go to the county fair and try to win a teddy bear by tossing a dime on a plate.

Betting on 7 is called a one-roll bet, because the next roll of the dice determines the result of the bet. There are many other one-roll bets. For example, by placing chips in the appropriate box in the center of the betting layout, you can bet on any of the individual numbers 2, 3, 11, or 12. The house edge for each of these bets is 16.7 percent, the same as betting on 7. Surely, we can do better than this.

Any Craps Bet

The word *craps* is slang for 2, 3, or 12. You can bet that craps will be the next roll of the dice by placing chips in the ANY CRAPS box in the center of the betting layout. If 2, 3, or 12 comes up on the next roll, you win. If anything else comes up, you lose. The house edge for this bet is 11.1 percent, not as bad as betting on individual numbers but about twice as bad as roulette.

Field Bet

The field bet is another one-roll bet. You make it by placing chips in the FIELD box on the betting layout. If the result of the next roll is 3, 4, 9, 10, or 11, you win and are paid even money (1 to 1 payoff odds). If 2 or 12 come up, you are paid double (2 to 1). If the result is 5, 6, 7, or 8, you lose. The field bet has a house edge of 5.6 percent, similar to roulette bets. Ho hum.

Finally, Something Better (in a Way) than Roulette: The Pass-Line Bet

The pass-line bet is the central bet of craps. When a sequence ends (winners paid off, losers' chips collected), the next sequence begins; dice rolls at the craps table can be thought of as never-ending pass-line betting sequences. When you see a crowded craps table with bettors screaming for the shooter to make the point, they are referring to the pass-line bet. All other bets are sometimes referred to as side bets.

Here's how this complicated bet works:

You place chips in the PASS LINE area of the betting layout. The first roll of a pass-line betting sequence is called the come-out roll. Three things can happen on the come-out roll:

7 or 11 comes up. You win.

2, 3, or 12 comes up. You lose.

4, 5, 6, 8, 9, or 10 comes up. These numbers are called *points*. When a point comes up on the come-out roll, the shooter continues to roll the dice until one of two things happens: Either the point is matched before 7 comes up, in which case you win, or 7 comes up before the point is matched, in which case you lose.

Dice rolls that take place after the come-out roll until either the point or 7 come up are called point rolls. During point rolls the only outcomes that affect the pass-line bet are the point and 7. This is the main thing: You want the point to come up before 7. (The results 2, 3, and 12, which cause you to lose on the come-out roll, and 11, which causes you to win on the come-out roll, don't count on point rolls.)

For example, suppose the come-out roll (first roll of the sequence) is 8. The point then becomes 8. For you to win your pass-line bet, 8 must be rolled before 7. Suppose the shooter then makes the following point rolls: 9, 3, 11, 4, 5, and 8. Since 8 came up before 7, you win. For another example, suppose the come-out roll and corresponding point is 4. The point rolls that follow are 5, 6, 12, 9, 9, and 7. Since 7 came up before the point of 4, you lose.

As soon as a pass-line sequence is ended, the next roll of the dice is another come-out roll. And so on. If a pass-line betting sequence ends with 7 on a point roll (you lose), it is customary for the shooter to pass the dice to the next player, who then becomes the new shooter.

Payoff odds for the pass-line bet are 1 to 1. The chance of winning, rounded off to three decimal places, is .493—almost (but not quite) 50-50. If you persistently play the pass line, on the average you will win 493 and lose 507 bets in 1,000, for a net loss of 14 bets per thousand. This yields a house edge of 1.4 percent, considerably better than the 5.3 percent house edge in roulette. In fact, the pass-line bet has one of the lowest house edges in the casino.

Casinos are always on the lookout for dice cheats trying to make a quick score. A popular con is for the cheat to secretly conceal "loaded" dice in his palm. Loaded dice are dice in which small weights, or loads, are inserted to make some sides more likely to come up than others. A variation of loaded dice is to shave a few millimeters off certain sides. This has the same effect as loading the dice. When it's time to be the shooter, the cheat subtly substitutes the palmed, loaded dice for the fair, casino dice, plays the game, and then replaces the trick dice with the fair dice before leaving the table with his accumulated winnings. Some cheats are quite good at this maneuver and manage to escape detection by the casino; however, most are eventually apprehended.

I heard a story about a cheat who was playing craps in a small casino. When it came time to be the shooter, the cheat placed a large bet on the pass line, picked up the casino dice, and smoothly switched them for the palmed, loaded dice. Unfortunately, his palms were sweaty, and when he rolled the loaded dice, one of the newly palmed casino dice came loose, causing three dice to bounce across the table, all landing on 5. Undaunted, the pit boss stared coldly at the cheat and said, "The point is fifteen."

Is the pass-line bet really better than roulette? While it is true that the 1.4 percent house edge for pass-line bets is better than the 5.3 percent house edge for roulette, don't start celebrating yet. In the long run the persistent pass-line bettor will still be a loser. Also, craps is a faster-moving game than roulette. In an hour of play, the pass-line bettor will make many more bets than the roulette bettor. If we use the "lowest hourly loss" criterion, the pass-line bettor doesn't do much better than the roulette bettor. Too bad.

The Come Bet

The trouble with the pass-line bet is that you can only make it on a come-out roll. During sequences of point rolls, there's nothing to do but admonish the gods to be nice. This is not good. The bettors like action, and the casino doesn't like bettors who aren't betting. The come bet allows the bettor to make the equivalent of a pass-line bet on every roll of the dice.

If the current roll is a point roll, and you want to make a pass-line bet, place chips in the COME area of the betting layout. When you make a come bet, for you the next roll of the dice is the same as the come-out roll on a pass-line bet. Since come bets are identical to pass-line bets, the house edge is the same: 1.4 percent.

Betting with the Casino (Almost)
The Don't-Pass Bet

The don't-pass bet is the opposite, almost, of the pass-line bet and allows a wily bettor to bet with the casino, almost. It's the *almost* that prevents the don't-pass bettor from having a winning proposition.

You make a don't-pass bet by placing chips in the DON'T PASS area of the betting layout. This bet, with 1 to 1 payoff odds, can be made only on a come-out roll and is determined as follows:

Come-out roll

7 or 11: lose
2 or 3 : win
12 : tie (in some casinos, 2 rather than 12 is a tie)

Point roll

....

point before 7 : lose
7 before point : win

....

Thus, don't-pass bets win whenever pass-line bets lose, except when 12 comes up on the come-out roll, in which case the bet is a tie (no money changes hands). The reason for the tie is simple: so that the casino can still have an edge. When the smoke clears, the house edge for don't-pass bets (rounded off to three decimal places) is the same as for pass-line bets: 1.4 percent. And all because of that pesky tie, when 12 comes up on the come-out roll.

The Don't-Come Bet

The don't-come bet is the same as the don't-pass bet, except it is made on a point roll. In other words, the don't-come bet is to the don't-pass bet as the come bet is to the pass-line bet. The don't-come bet allows you to make a bet equivalent to the don't-pass bet on every roll. Since it's equivalent to the don't-pass bet, the house edge for don't-come bets is 1.4 percent. Whew.

The "Simplicity of Betting" Rule for Craps

The bad house edge (16.7 percent) for one-roll bets, compared with the not-so-bad house edge (1.4 percent) for the pass-line bet and its counterparts, illustrates the simplicity-of-betting rule for craps: *The simpler the bet, the bigger the house edge.*

Summary

Craps, like roulette, is a game of pure chance. No skill involved. There are a variety of bets at the craps table, the best of which are the pass-line bet and its counterparts, for which the house edge is 1.4 percent. While it is true that the persistent bettor will eventually go broke, the pass-line bet offers the prudent gambler an opportunity to spend an enjoyable evening playing an exciting game, with a slight chance of going home a winner.

Chapter 7 🎲 🎲

Slot Machines and Video Poker: Pavlov Lives On

"Suckers have no business with money anyway."
CANADA BILL JONES, LEGENDARY THREE-CARD MONTE DEALER

Slot Machines

If you want to go into a hypnotic state and have a chance to win some cash as well, look no farther than the slot machine. Sit on a comfortable stool, insert money, tokens, or a special slot-machine credit card into the machine, pull a handle or push a button, watch symbols fly by on a colorful video screen, and hope that the configuration that finally appears on the center line is a winner. If you are so lucky as to win, bells and whistles sound, and coins clank into the machine's payoff bin for those around you to see. As you collect your winnings, you understand that your patience has paid off. Unlike the clowns nearby, you have been chosen by the fates to be the lucky one. This may be an omen that portends future wealth. You put more money in the machine and become one with the cosmos, perhaps dimly aware that the excitement you get from the bells and whistles, even when you're not a winner, puts you into the same state of mind as Pavlov's drooling dogs. When you eventually lose the allotted amount for the evening, you stagger over to the gift shop and numbly check the price of T-shirts and teddy bears.

Slot machines have been around for about a century. Your basic machine is a mechanical device that has three or four internal "reels" of symbols, such as fruits, genies, dollar signs, or oil wells. You insert coins and pull the machine's handle or arm (hence the term "one-armed bandit"), activating a mechanism that causes the reels to spin. When the reels come to rest, you get a payoff if specified configurations of symbols appear on the center line in the display window on the front of the machine (for example, if all symbols on the center line are cherries). The winning configurations and resulting payoffs are displayed on a payoff table posted on the front of the machine. Mechanical slot machines still exist in many casinos, but computer technology has given rise to video slot machines.

Video slot machines and video poker machines are casinos' biggest profit makers. A video slot machine is activated when the player inserts a coin and pulls a handle or presses a button. A random-number generator in the machine's computer program picks an outcome and, after an appropriate rendition of sound effects and video graphics, displays a configuration of symbols on the machine's video monitor. Just like a mechanical slot machine, the player wins if the randomly selected outcome is a winning configuration, as specified by the machine's payoff table.

A slot-machine player has no choices during the course of play. Thus, there are no slot-machine playing strategies. The game is pure chance. If you win, you got lucky. Different slot machines have different house edges, so the only real choice available to the slot-machine player is to decide which machine to play—or whether to play at all. In this sense, an optimal slot-machine strategy is to find the machine with the best house edge. The only way to determine the house edge for a video slot machine without knowing something about the machine's software is to do a Monte Carlo simulation: Play the machine many times, always betting the minimum denomination, and see how many plays it takes for you to go broke. If you divide the number of plays it takes to lose your money into your starting bankroll and divide that result by the denomination of the machine, you'll get an estimate of the house edge.

For example, suppose you're playing a quarter machine and that you start with $50. Suppose it takes 430 plays until your $50 is gone. Dividing 50 by 430 gives an average loss of $.1163 per play. Since this is a quarter machine, divide the amount by 25 cents, yielding $.4651. In other words, you lost at the rate of 46.51 cents per dollar bet, giving an estimated house edge of 46.51 percent. In general, slot-machine house edges range between 2 percent and 50 percent.

The house edge on most video slot machines can be changed at the whim of the casino by flipping a switch or with software that controls the game program. Too many winners? No problem. The casino can fix that. Too many losers? That can be fixed as well. Too many casino employees working at the craps table? Shut down the craps table and install more slot machines. Nobody has to work at the slot machines, except waitresses bringing free drinks, change makers, security guards, and the maintenance crew. That's one of the reasons they're so profitable. The other reason has to do with the last word in the term "one-armed bandit."

Video Poker

Video poker allows you to play a computerized version of draw poker (or some other type of poker) and bet on the outcome. Here is a general summary of a video draw-poker game:

The player inserts money or a token into the machine. A random-number generator in the machine's computer program simulates the deal of an initial, five-card hand from a well-shuffled deck. The cards are displayed on the machine's video screen. By pushing buttons or using a touch screen, the player indicates which cards to hold and which to discard. In the same way, the player signals the computer to randomly select replacement cards for discarded cards and display them on the video screen. The player wins if the final poker hand is a winning hand, as indicated by game's payoff table, which is a statement of the winning poker hands and payoffs per unit bet for those hands.

There are important differences between regular draw poker and video draw poker:

- Regular draw-poker players use psychological skill factors, such as bluffing and reading an opponent's mannerisms, to gain an advantage. You can neither bluff nor read the mannerisms of a machine.

- In regular draw poker there is a round of betting after the initial hand is dealt and again after the draw. In video draw poker, the only betting is at the beginning of the game, before the machine deals the hand.

- In regular draw poker there are two or more players competing against each other. The player with the best poker hand wins the pot. In video draw poker, there is only one player, playing against the machine. The video player only wins if the final poker hand is on the game's payoff table.

- In regular draw poker a player can fold after seeing the initial hand, thereby not having to bet on that hand. In video draw poker a player must make a bet before play begins. The only folding in video draw poker is when the change person folds your dollars and gives you more coins.

- In video draw poker a player pays and is paid by the casino, not other players. In ordinary poker games, the losing players pay the winning players. The casino makes money by taking a percentage of each pot or by charging players a fee to play.

Even skilled, ordinary poker players have trouble making a buck, as the following cautionary tale shows. A professional poker player vacationing in Montana stopped at a local tavern to have a beer. A poker game was in progress, so the gambler decided to take the local cowboys for a few dollars. After playing for a while, the gambler was dealt a flush. The betting was heavy. Soon there was $5,000 in the pot and one cowboy left betting against the gambler. When the players showed their hands, the cowboy had 2, 4, 6, 8, 10, two hearts, three clubs. The gambler laughed, showed his flush, and started collecting the pot. The cowboy stopped him. "You lose," said the cowboy. "I've got 2, 4, 6, 8, 10, two hearts, three clubs. That's a Lollapalooza. In this town, a Lollapalooza beats anything."

Looking around the room, the gambler decided not to argue with the cowboys. He figured he'd recoup his losses, which he did in an hour. Then, the gambler was dealt 2, 4, 6, 8, 10, two hearts, three clubs—a Lollapalooza. The gambler bet big. Soon, there was $20,000 in the pot and one cowboy left in the hand. The cowboy showed a full house. "Sorry," said the gambler, as he started raking in the pot. "This time, I've got the Lollapalooza."

"Not so fast, podner," said the cowboy as he grabbed the gambler's arm. "Only one Lollapalooza per night!"

Playing Strategies

In a video draw-poker game the computer randomly selects both the initial hand and replacement cards, simulating ordinary draw-poker play. A player's only choices during the course of play are deciding which cards to hold and which to discard for a given initial hand. Thus, a playing strategy is a rule that specifies such choices for any given initial hand.

For games in which there are no playing strategies, such as roulette, craps, and slot machines, there is only one house edge for each bet. As we have seen, the house edge for most roulette bets is 5.3 percent, whereas the house edge for the pass-line bet in craps is 1.4 percent. As I mentioned earlier, house edges for slot machines vary from machine to machine, typically ranging from 5 percent to 50 percent. The house edge for a video draw-poker game depends not only on the payoff odds on the game's payoff table but also on the player's strategy, namely, the various hold/discard choices for a given initial hand.

A player's optimal video draw-poker strategy is that which yields the lowest house edge (or highest expected payoff). If the machine manufacturer wants to guarantee a long-run profit for the machine's owner, the game's optimal strategy must have a positive house edge. This can be accomplished with an appropriate payoff table. In this case, even a player using the optimal strategy will lose in the long run.

Is video draw poker a game of chance or a game of skill? There are various ways to define skill. According to the *Random House Webster's Unabridged Dictionary*, skill is "the ability, coming from one's knowledge, practice, aptitude, etc., to do something well." What this means depends on the context. For a video draw-poker game, since a player's only choices are which cards to hold and discard for a given initial hand, with everything else decided by chance, the natural measure of skill is the house edge for a particular strategy. A player who uses the optimal strategy for a particular game would be the most skillful player of that game. This definition of skill has its limits, since even a player using an optimal video draw-poker strategy will lose money in the long run if the house edge is positive, a factor that can be controlled by modifying the pay-off table on the machine.

Computing Video Draw Poker Probabilities

Computing probabilities for video draw poker can be a tedious task. For example, suppose you are dealt the 10 of clubs, queen of clubs, jack of clubs, jack of hearts, and jack of spades. Should you keep the three jacks, discard the 10 and queen, and hope for four jacks or a full house? (You already have three jacks.) Or should you forget about the three jacks and discard the jack of hearts and jack of spades to go for the straight flush or royal flush?

Suppose that you decide to discard the 10 and queen. What is the probability that your final hand is three jacks? Four jacks? A full house?

Since five cards have been dealt, there are 47 cards left in the deck. The problem then becomes computing probabilities for drawing two cards at random from a deck containing the 47 remaining cards. To compute such probabilities, you must know how to count combinations. I'll spare you the details. It turns out that there are 1,081 different combinations of drawing two cards from 47 and that the resulting probabilities for the corresponding final poker hands are as follows:

Four of a kind: $46/1{,}081 = .04255$
Full house: $66/1{,}081 = .06105$
Three jacks: $969/1{,}081 = .89639$ (no improvement
                                          from the draw)

On the other hand, if you discard the jack of hearts and jack of spades to go for the royal flush or straight flush, you have the following probabilities of succeeding:

Royal flush (king and ace of clubs): 1/1081 = .00093

Straight flush (8 and 9 of clubs, or 9 and king of clubs): 2/1081 = .00185

There are some other, lesser payoff possibilities for this draw as well (flush, straight, three of a kind, two pair, pair), but the high-payoff outcomes are the straight flush and royal flush. In any event, once all these probabilities are computed, you must factor in the payoff odds to see whether it makes sense to use the nonintuitive strategy of giving up three jacks to go for a royal flush or straight flush. This may be the best thing to do in a progressive game with a big royal flush payoff.

In order to put together an optimal video draw poker strategy, you must do this type of computation for every possible initial hand. If you are determined to play this game, doing some Monte Carlo simulations or reading Stanford Wong's book are sensible alternatives.

Progressive Games and Winning Strategies

Along with the popularity of video gambling machines has come the proliferation of *progressive* games. In a progressive game, a group of video slot machines or video poker machines are networked together and controlled by a central computer. As long as nobody wins, a few cents from each bet are added to the jackpot. (In the case of video draw poker, the jackpot is typically a royal flush.) The progressive payoff accumulates until there is a winner. The machines sharing the progressive jackpot may be clustered together in the same casino or, thanks to computer technology, spread out over many casinos, possibly in different cities. Since the machines are programmed so that the chance of winning the jackpot is very low, large progressive jackpots, sometimes running into the millions, can result.

One of the reasons casinos started offering progressive games was to compete with state lotteries, whose multimillion-dollar jackpots were thought to be luring gamblers away from casinos. Now gamblers who want to become instant millionaires need not rely on boring state lotteries that don't offer the video special effects, free drinks, and other comps lavished by casinos on slot-machine players.

Out of the muck a winning game emerges. An interesting feature of progressive video gambling payoffs is that, unlike state lotteries, they have no multiple jackpot winners. Whoever hits the jackpot first gets all the money. This means that when the jackpot in a progressive game gets high enough, a player can have a positive expected payoff, in other words, a winning game. This fact has not escaped the attention of the clever gamblers who lurk around casinos looking for good bets the way vultures lurk around the interstate looking for roadkill. In recent years, because of the huge jackpots and the occasional edge for the gambler, progressive games have achieved well-deserved renown, with patient players spending many hours engaged in progressive play. Unfortunately, there is a problem.

Although the progressive jackpot may get large enough to offer a positive expected payoff, if you are continuously playing a progressive machine and another machine in the network is left unattended, some lucky clown may come up and win *your* jackpot. In fact, although your chances of getting particular outcomes remains the same, just one other player in the progressive network cuts your chance of winning the jackpot payoff in half: If the other player hits the jackpot first, you lose. If you are playing one of ten machines in a progressive network, and other players start betting at all the other machines, your chance of winning the progressive jackpot becomes one-tenth what it was when you were the only player, and your positive expected payoff vanishes like a puff of smoke. Fortunately, there is a partial solution to this dilemma: team play.

Progressive Video Gambling Teams

Slot-machine teams are important in the quest for the big, progressive jackpot because it takes a group of players working together to commandeer an entire network of machines, thus insuring that no outsider will win the progressive jackpot, at least while the team has control of all the machines. Since some progressive networks involve machines in many casinos, it may be necessary for a slot-machine team to spread out over a wide area.

If the chance of winning is low, plan to play for a while. Even with team play, a difficulty in winning progressive jackpots arises from the lottery principle: Since the chance of winning is so low, it may take a large investment of time and money to insure a win. For example, if the chance of winning the jackpot for a $1 progressive slot machine is one in a million and the jackpot goes high enough above $1 million, you and your teammates have a positive expected payoff; however, it will take an average of a million plays for the big payoff. Suppose you make five plays per minute, and suppose also that there are ten machines in the network. Then it will take you and nine teammates two weeks of continuous play (twenty-four hours a day) to make a million plays. During this time you will win various lesser payoffs, but a considerable starting bankroll will usually be necessary for survival. Also, since you now have ten people playing to win one jackpot, if you or one of your compatriots wins, you will have to split the pot ten ways. This is similar to what happens with a lottery jackpot.

I heard about a team of progressive slot-machine players who descended on a cluster of machines in a Nevada casino when the jackpot got high enough to give them a positive game. They played nonstop for forty-eight hours, afraid to relinquish a machine for fear that someone else would win the jackpot. Exhausted, they finally gave up with no jackpot win, after losing thousands of dollars. A few hours later a retired couple driving across the country in a beat-up Winnebago stopped at the casino for a break and won the jackpot after putting $5 in one of the machines. Surprisingly, the tired slot-machine team wasn't particularly upset about the tourists' lucky win. They knew that as long as they stuck to games with positive expected payoff, they were doing the right thing, at least theoretically.

Chapter 8 🎲🎲

Blackjack: A Game You (or Someone) Can Win

"Grain upon grain, one by one, and one day, suddenly, there's a heap, a little heap, the impossible heap."
SAMUEL BECKETT, *ENDGAME*

*B*lackjack is one of the few casino games in which the player can have a mathematical edge over the casino. Winning blackjack strategies were developed by statisticians doing computer simulations in the 1950s and publicized in Edward O. Thorp's *Beat the Dealer,* which sent gamblers by the thousands flocking to casinos to play blackjack. Fortunately for the casinos, winning strategies were difficult to master and gave the player only a slight edge under ideal playing conditions. Even so, casinos, temples of chance though they are, didn't want to take any chances. Multiple-deck games were introduced. Rules were changed. Good players were asked to leave.

Suppose you master a winning blackjack strategy. Assuming perfect playing conditions (single- or double-deck game, with a dealer who deals through the deck and doesn't cheat), you have about a 2 percent edge. How much you win in the long run depends on how much you bet. If you bet $50 per hand and play one hand per minute, you'll bet $3,000 per hour. With a 2 percent edge, your hourly winnings will average $60. This doesn't take into account travel expenses, time at the casino when you're not playing, and finding new places to play after you get thrown out of the old ones.

Blackjack Basics

Blackjack, or 21, is a card game played between the dealer (a casino employee) and one or more players. The object of the game is to get a score higher than the dealer's score without exceeding 21, the best possible score.

Your score is the sum of the values of each card in your hand. Cards 2 through 9 are worth face value. Tens, jacks, queens, and kings are worth 10 points, and aces are worth either 1 or 11.

The game is played at a table with a betting layout. Before any cards are dealt, you bet by placing chips in the circle in front of you on the betting layout. After bets are made, the dealer deals two cards to each player. One of the dealer's cards is dealt face up ("up card") and the other face-down ("hole card"). Your first two cards are sometimes dealt face up, sometimes facedown, and sometimes one up and one down. Since the dealer has to play by predetermined rules, it doesn't really matter how your cards are dealt, unless you are keeping track of the cards as they are dealt from the deck, in which case it will be easier to see players' cards if they are all dealt face up.

After two cards are dealt to everyone, play begins with the first player to the left of the dealer. There are two main choices: "stand," which means to take no additional cards, and "draw," which means to take an additional card. To stand, you slide your cards under your chips. To draw, you scratch your cards toward yourself on the table. If you draw, the dealer deals you a card, face up. After looking at your new card, you again decide whether to stand or draw, and so on. When you finally stand, your hand is complete, and play moves to the player on your left.

If your score ever exceeds 21, you "bust" and must turn your cards face up. When you bust, you lose.

When all the bettors are finished playing, if any haven't busted, the dealer plays. Unlike the players, the dealer must draw with a score less than 17 and stand with a score of 17 or more. If the dealer busts, any players who haven't busted win. If the dealer doesn't bust, players with a score higher than the dealer's score win and those with a lower score lose.

Players with the same score as the dealer tie, and no chips change hands. Because players go first, if you bust, you lose, even if the dealer busts later. Payoff odds on regular hands are 1 to 1.

If your hand contains an ace that can be counted as 11 points without the score exceeding 21, it is called a "soft" hand. Otherwise, it is called a "hard" hand. You compute the score of a soft hand by counting the ace as 11. For example, the hand 5, 3, ace is "soft 19." The hand 7, 9, ace is "hard 17" because if you count the ace as 11, the score exceeds 21. You can't bust by drawing to a soft hand. This doesn't mean that you should draw to a soft hand. For example, the hand 4, 5, ace is "soft 20," a good hand. In many casinos, the dealer must draw with soft 17.

After being dealt the first two cards, you have the option of "doubling down," that is, doubling your bet and receiving exactly one more card. To do this, you turn your cards face up and place an equal amount of chips next to your original bet. The dealer then deals you a card facedown. Your hand consists of these three cards and will be compared with the dealer's hand after the dealer plays. The bet is twice your original bet.

Doubling down is smart if there's a good chance you will beat the dealer with one more card. For example, if your hand is 7, 4, there's a good chance you'll be dealt a 10-value card, giving you 21.

The rules for doubling down vary: In some casinos you can double down with any two initial cards; in others, you can double down only with a score of 10 or 11.

If your first two cards are the same denomination, you can turn your hand into two hands by "splitting pairs." For example, suppose you are dealt a pair of 8's, giving you a score of 16. To split the pair, turn the 8's face up, putting one in front of your bet and the other next to it. Put chips equal to your original bet next to the second 8, and tell the dealer you are splitting the pair. The dealer will treat your pair of 8's as the beginning of separate hands, dealing to them one at a time, face up.

If yours or the dealer's first two cards are an ace and a 10-value card (10, jack, queen, king), the hand is called a "natural" or "blackjack." If you have a natural and the dealer doesn't, you win. The payoff odds for a natural are 1.5 to 1.

After the first two cards are dealt, if the dealer's up card is an ace, you can make an "insurance" bet that the dealer has a natural by placing up to half your original bet in the "insurance" area on the table. The dealer asks the players if they want to make this bet. Payoff odds for an insurance bet are 2 to 1. This bet is supposed to give the player "insurance" against losing to a dealer's natural.

After insurance bets are made, the dealer peeks at his hole card. If he has a natural, insurance bets win, but the dealer's natural beats everyone who doesn't have a natural. Thus, if you make an insurance bet and don't have a natural, and the dealer has a natural, you win your insurance bet and lose your original bet. If the dealer doesn't have a natural, you lose your insurance bet and play out the hand. If the dealer and you both have naturals, you win your insurance bet and tie your original bet. And so it goes.

As with any bet, you shouldn't make an insurance bet unless you have an edge over the house. You win an insurance bet when the dealer's hole card is a 10-value card. In a full single deck, 16 cards are 10-value cards. If the dealer's up card is an ace, 16 of the remaining 51 cards are 10-value cards, so, disregarding other cards you may have seen, the chance that the dealer's hole card is a 10-value card equals 16/51, or .31. The payoff odds on an insurance bet are 2 to 1, so your chance of winning has to be greater than 1/3 = .33 for the bet to be favorable. Thus, you shouldn't make an insurance bet when the deck is full, since the fraction of 10-value cards equals .31, less than .33. In general, don't make an insurance bet unless you know that the fraction of 10-value cards remaining in the deck is greater than 1/3.

Winning Strategies

Computer simulations in the 1950s produced blackjack strategies that give the player a slight edge over the casino. Winning strategies are based on the fact that the dealer must draw with a score of 16 or less, whereas a player can draw or stand with any score. The player's advantage comes from situations where the dealer is "forced" to bust. For example, a dealer with 3, 4, 5, or 6 showing is forced to draw with a 10-value card in the hole. In such a situation, the player should be conservative about drawing so as not to bust first. When the dealer has a high card showing, a player should be bolder about drawing.

A player's chances of winning can be increased by keeping track of cards as they are dealt. Card counting, as this is called, requires a particular type of skill, namely, the ability to keep mental track of a rapidly changing tally in a tumultuous casino environment. Roughly speaking, when the deck is rich in 10-value cards, there is a greater chance that the dealer will bust, and so the player should make relatively large bets and be conservative about drawing. When the deck is rich in low-value cards, there is less chance that the dealer will bust, and the player should make smaller bets. It takes concentration and dedication to master the task of card counting to the point where you actually have an edge over the casino. Even then, you may invest so much time in earning a long-run profit that your hourly wage will become a pittance unless you make large bets.

Card Counters Versus the Casino

When *Beat the Dealer* was published, panicked casinos tried to counter the counters in various ways, such as ejecting them from the casinos. With the aid of one-way mirrors above the tables (the "eye in the sky"), photos were taken to identify the professionals. Card counting isn't illegal, but who wants to argue legalities with a security person who asks you to leave the premises?

Card counters responded to the casino crackdown in various ways. One bettor I know who had been thrown out of every big casino in Nevada decided to change his appearance. Sheldon, a pasty-faced gambler who went outdoors only to go from one casino to another, grew a beard, dyed his hair, went to a tanning salon, and bought a flashy suit. He tested his disguise on his friends and family, and his brother didn't recognize him. He went to a Las Vegas casino and sat down at a blackjack table. Five minutes later, the pit boss approached him and said, "Hi, Sheldon." When Sheldon asked the pit boss how he recognized him, the pit boss pointed to his watch. Sheldon had forgotten to remove the monogrammed watch his girlfriend had given him a few years earlier.

Since winning blackjack systems require the player to make big bets when the deck is "good" and small bets when the deck is "bad," casinos sometimes try to spot card counters by watching for large bet variations. A typical tourist won't make a sequence of $5 bets followed by a few $100 bets, then go back to $5 bets. One gambler I know was kicked out of a Lake Tahoe casino merely for varying his bets. He had lost $1,000 at the time he was ejected. Another professional blackjack player, Lenny, invented an elaborate strategy for varying his bets without getting caught. He called his system the "big player" concept. Lenny organized three teams of blackjack players, with five players on a team. Each team consisted of four card counters and a big player. The team would enter a casino one by one, the card counters sitting at separate blackjack tables. The counters counted cards and made minimum bets. Meanwhile, the big player swaggered through the casino like a drunken high roller. When the count at a table got high, the card counter at this table would signal the big player, who would lurch over and make big bets. If the count got low, the counter would signal again, and the big player would leave the table. The counters were always making small bets, and the big player was always making big bets: no bet variations. It took a year for the casinos to catch on. When they did, Lenny and his teams had to leave town in a hurry.

When Lenny's big-player teams were disbanded, he hit on another scheme for beating the casinos. He hired a computer expert to develop a card-counting computer small enough to be taken into a casino. The computer was strapped around the bettor's thigh. Small sensing devices were placed under the bettor's toes for tapping information into the computer as cards were dealt, and a buzzer signaled the bettor to draw or stand and how many units to bet. Lenny's computer was a failure. Although it worked in the lab, people had trouble using it in a casino environment. Lenny later tried hiding a similar device in a pair of large sunglasses. A pit boss soon booted him out.

An Almost Winning Strategy

This strategy is a variation of the basic strategy given in most blackjack books. There is no card counting. All you do is memorize these simple rules for how to play, given your current total and the dealer's up card:

Drawing and Standing

1. Draw with a hard hand of 11 or less.
2. Draw if you have a hard hand with a score between 12 and 16 and the dealer's up card is 7 or higher, including ace.
3. Draw if you have a soft hand of 17 or less.
4. Draw if you have soft 18 and the dealer's up card is a 9, 10, or ace.
5. Stand if none of these conditions are met.

Doubling Down: Takes Precedence over Drawing.

1. Double down if your score is 11.
2. Double down if your score is 10 and the dealer's up card is 9 or less.
3. Double down if your score is 9 and the dealer's up card is 3 through 6.
4. Double down if your first two cards are an ace and a 2 through 7 and the dealer's up card is a 4, 5, or 6.

Splitting Pairs: Splitting Pairs Takes Precedence over Drawing.

1. Always split a pair of aces or 8's.
2. Never split a pair of 4's, 5's, or 10's.
3. Split a pair of 2's, 3's, 6's, or 7's when the dealer's up card is 3 through 7.
4. Split a pair of 9's when the dealer's up card is 2 through 6, 8, or 9.

Insurance

Never buy insurance unless the fraction of 10-value cards in the deck is less than 1/3. Unless you're counting cards and can make this determination, don't buy insurance.

Developing a Good Blackjack Strategy

Finding basic blackjack probabilities isn't extremely difficult. The hard part is putting them all together to form a
good strategy. For example, suppose you are playing in a
single-deck game, are dealt an 8 and a 7, and the dealer's up
card is 9. Suppose also that you haven't seen any other
cards. What is the probability that you will bust if you draw
one more card?

With a total of 15, you will bust on the next card only if you
draw a 7, 8, 9, 10, or face card. There are four cards of each
denomination, but one 7, 8, and 9 have already been dealt.
Thus, there are three 7's, 8's, and 9's remaining in the deck,
along with four 10's, jacks, queen, and kings, for a total of
25 cards that will cause you to bust. Since there are 52 cards in
the deck, of which you have seen 3, there are 49 cards remaining, so the probability that you bust on the next card is 25/49.
(Other cards that may have been dealt but that you haven't
seen don't affect this computation.) It follows that the probability that you don't bust is 24/49, slightly less than a 50 percent chance.

Knowing this probability isn't enough to determine
whether to stand or draw with a 15 against the dealer's 9. For
example, you need to find the chance that the dealer, with an
up card of 9, will end up with various totals. This is tedious,
because you must consider every possible hole card. Once you
have done this, you can then compute the probability that the
dealer will win if you stand with a 15. Then, after finding your
chances of achieving various totals, should you draw with a 15,
you can decide whether to stand or draw. (According to the
basic strategy, you should draw.) In order to develop a good,
comprehensive strategy, you must do this for every possible
total in your hand for every possible up card that the dealer
might have. This is a daunting task, which is why Thorp and
others resorted to computer simulations to get answers.

Does It Matter Where You Sit at the Table?

Many blackjack players believe that a player's location at the
table is an important factor in blackjack play. For example,

some gamblers prefer to sit to the immediate right of the dealer (from the dealer's perspective); this seat is called third base, because, since the dealer deals from left to right, the player at third base is the last player dealt to. Other players prefer to sit at first base, believing that since they are the first players dealt to, they are more likely to get good hands.

The truth is that if the cards are well-shuffled, your overall chances of getting dealt particular cards are the same, no matter where you sit at the table. Of course, if you see the player sitting next to you get the 3 of hearts, you know that you won't be getting that card, however, at the time the deck was shuffled, everyone had the same chance of getting it.

If you are using a card-counting strategy or somehow basing your play on which cards have been dealt from the deck, sitting at third base provides a small benefit: Since you are the last player to play, you get to see more cards before you must decide what to do. If you are using a basic strategy or not basing your play on which cards have come up, it makes no difference where you sit at the table.

Does It Matter How Many People Are Playing?

Some players prefer to be the only person at the table, while others are more comfortable sitting in a crowd. Players who know the rudiments of good blackjack play may get annoyed at players who make dumb decisions. It may also be annoying to have smoke blown in your face or to sit next to a boisterous drunk who keeps you from concentrating on the game.

From a mathematical perspective, if you are using a basic strategy or not keeping track of cards as they are dealt from the deck, it doesn't matter how many people are playing at your table. If you are a card counter, and the dealer deals down to near the end of the deck (or decks), it may be advantageous to be the only player at the table. This is because when the deck is favorable to the player, there may be more opportunities to make large bets when there aren't other players to eat up the remaining cards. On the other hand, if you are a card counter worried about being spotted by the casino, you may prefer the relative anonymity of playing at a crowded table of noisy tourists.

Does It Matter How Many Decks Are Being Used?

If you are a card counter, games with just a single deck or double deck offer more situations that are extremely favorable to the player, providing that the dealer deals to near the end of the deck(s). On the other hand, most blackjack games in today's casinos are multiple-deck games dealt from a plastic shoe designed to hold many decks of cards. Thus, it may be more convenient to play in a multiple-deck game. In order to avoid the long periods of play with unfavorable cards that occur in multiple-deck games, some card counters lurk around a table, keeping track of cards as they are dealt from the deck, waiting for the deck to get favorable before jumping in and making a bet. This can be risky, because in a crowded casino the table may be full when it's time to make your move. If you are not a card counter, it makes no difference where you play. You should gamble at whichever table makes you happy.

Another factor you may want to consider is the possibility of being cheated. It's difficult for the dealer to cheat in a game dealt out of a shoe, unless the shoe itself is rigged, whereas with practice most dealers can master sleight-of-hand techniques that make it easy to cheat when holding the deck.

How Can You Tell If You're Being Cheated?

Thorp and other authors talk about dealers who cheat, either on their own or as part of the casino's attempts to get card counters to leave. The cheating is presumably carried out in various ways, usually involving the same sleight-of-hand techniques that magicians use when doing card tricks. For example, *dealing seconds* involves dealing the second card from the top of the deck instead of the top card, keeping the (known) top card ready for use when the dealer needs it. It may seem unlikely that a dealer could get away with this when a group of gamblers are sitting a few feet away, transfixed on the cards, however, with practice this technique can be essentially undetectable, even under close scrutiny. Other cheating techniques include the following:

- Trick shuffles that keep high-value cards at the bottom of the deck, thus giving the dealer a big advantage
- Removing high cards from the deck before the deck is put in play
- Trick shoes that allow the dealer to secretly see the top card and deal seconds in multiple-deck games

It may seem unclear why dealers would want to cheat. What's in it for them? They don't get a percentage of the casino's winnings. Since casinos videotape games from the "eyes in the sky" located above every table, it is risky for a dealer to cheat on a regular basis. A dealer has two principle motivations for cheating: (1) If the dealer cheats to let an accomplice win, someone else must lose, since the casinos monitor the average take at every table; and (2) boredom.

There is probably not much cheating in today's corporate casinos, but it never hurts to be alert to the possibility. Here are four ways to be cautious:

1. Use your intuition. If you don't like the dealer, get up and move to another table.
2. If everyone at the table is losing (including you), move to another table.
3. Be statistical. If you are using the basic strategy or counting cards, the game is almost like a coin toss: In 100 bets there is a good chance you will win between 40 and 60 times. If you win more than 60, that's great, but if you win fewer than 40, go to another table.
4. Here's another way to be statistical: Keep track of the number of times you are dealt a blackjack. On the average, you should get about one blackjack every 20 hands. The chance is about 1 in 140 that you will not get a blackjack in 100 hands, so if 100 hands go by and you haven't gotten a blackjack, move to another table.

Remember, weird things happen just due to chance. Even if you use a good strategy, when you play for a long time you will have losing streaks as well as winning streaks. Professional blackjack players have many losing nights, even when they're playing under ideal conditions. Just because you lose doesn't mean that you're being cheated.

How Much Should You Bet?

If you are playing a basic strategy with no card counting, it is prudent to bet about 1 percent of your bankroll. For example, if you have $500, make $5 bets. If you have $1,000, make $10 bets. Since you are always betting the same percentage, if you win, your bet size will increase, and if you lose, your bet will decrease. This is called *fixed-fraction betting*: The more you have, the more you bet; the less you have, the less you bet. This runs contrary to the wacky notion of betting big when you lose, to cover your losses.

The Kelly System

In 1956 mathematician J. L. Kelly, Jr., published an article about mathematical-information theory applied to transmission of information over a phone line. The results were applied to gambling and became known as the Kelly system. Using the Kelly system, you always bet a *fixed fraction* of your bankroll (the total amount you have for betting purposes). Again, the more you have, the more you bet; the less you have, the less you bet. More precisely, you bet the fraction of your bankroll that maximizes its "growth rate" (growth rate is the expected logarithm of your return per dollar bet). The Kelly system applies only to situations where you have an edge. If the casino has an advantage, the Kelly system says you should go to the movies.

Statistician Leo Breiman showed that the Kelly system is optimal for two reasons. First, it will do better in the long run than any substantially different strategy. Second, the average number of bets necessary to reach a specific goal when using the Kelly system is lower than with any other strategy.

A third desirable property of the Kelly system is that since you always bet a fixed fraction of your bankroll, it's difficult to go broke. In fact, if money were "infinitely divisible" (it's not), you would always be able to bet a fraction of what you had left and would never go broke.

In general, Kelly-system fractions are difficult to compute. If the payoff odds for a single, win-lose bet (you either win or you lose) are 1 to 1, and p is the chance that you win, the Kelly system says that you should bet the fraction $2p - 1$ of your bankroll on each bet. For example, suppose a box contains 7 balls marked "win" and 3 balls marked "lose." A ball is drawn at random from the box. If a win ball is drawn, you win, and if a lose ball is drawn, you lose. Thus, your chance of winning this bet equals .7. Suppose that the payoff odds are 1 to 1. Nobody in his right mind would ever offer you this wager, but if someone does, the first thing to notice is that even though you should bet, you shouldn't bet your entire bankroll. Each time you make the bet, there is a 30 percent chance that you will lose, so if you wager your entire bankroll each time you bet, even though you have an edge, you will soon be broke. How much should you bet?

The Kelly system says to bet the fraction $2(.7) - 1 = .4$, or 40 percent of your bankroll on each bet. Thus, if you start with $1,000, your first bet should be $400. If you lose, your bankroll has decreased to $600, so your second bet should be $240. If you win, your bankroll has increased to $840, and your third bet should be $336, and so on. Contrary to popular betting strategies, in which you bet big to cover your losses, with the Kelly system, the more you have the more you bet and the less you have, the less you bet. This is crucial for an optimal betting system.

Suppose you make 100 bets in this game in this way, each time betting 40 percent of your bankroll. You will win an average of 70 and lose an average of 30 bets, causing your bankroll to increase by a factor of 3,745. If you start with $1,000 and make 100 bets with these results, you will end up with $3.75 million.

For most blackjack outcomes, the payoff odds are 1 to 1. If you are an expert card counter, playing under ideal conditions, your chance of winning a hand when the deck is in your favor may average about 51 percent (this will vary somewhat according to the composition of cards). Assuming a 51 percent win probability, the Kelly system says that you should bet roughly 2 percent of your bankroll. Thus, if you have a bankroll of $5,000, you should bet $100. If your bankroll is $1,000, you should bet $20. If your bankroll is $500, you should bet $10, and so on. This can be modified up or down, depending on the composition of cards remaining in the deck.

The Kelly System Applied to the Warped Roulette Wheel

In Chapter 5, I mentioned a pair of gamblers who searched Nevada for a warped roulette wheel, finally finding one after more than a year of hunting. I discussed the difficulties of ascertaining whether a wheel was warped by watching it in play, using as a hypothetical example a wheel in which there was a warp that increased the chance of 17 coming up to 2/38, instead of the usual 1/38. Suppose you discover such a wheel. How much should you bet? This is not an even-money wager, so the formula for computing the Kelly fraction has to be generalized, as follows: *For a single, win-lose bet, bet the fraction E/odds, where E = your expected payoff for a $1 bet, and odds = your payoff odds for a $1 bet.*

The hypothetical warped wheel has a 2/38 chance that 17 will come up. Since the payoff odds are 35 to 1, betting $1 each time, you will win $35 an average of 2 spins in 38, and lose $1 an average of 36 spins in 38, for a net profit of $70 − 36 = $34 in 38 spins. This gives you an expected payoff of $34/38 per spin. Since odds = 35 and E = 34/38, the Kelly fraction equals (34/38)/35 = .026. In other words, if you should be so fortunate as to discover such a warped wheel, you should always bet 2.6 percent of your bankroll: If you start with $10,000, bet $260. If you get unlucky and your bankroll decreases to $1,000, bet $26, and so on. Even though this may seem conservative, it is optimal: In repeated play, your bankroll will grow faster than with any other strategy.

The Kelly System Applied to the Stock Market

In order to find Kelly-system betting fractions you must know probabilities of winning as well as appropriate payoff odds. In the casino this information is usually obtainable, either by direct computations or, as in blackjack, with computer simulations. Unfortunately, the Kelly system only applies when the player has an edge, and there aren't many casino wagers of this type. The most popular gambling game outside the casino is the stock market. But probabilities that an investment will be profitable, as well as appropriate payoffs, are difficult to obtain and are typically estimated by the use of historical data, along with current market information. When these estimates are accurate, the Kelly system will yield optimal portfolio performance. In situations where there are a variety of investment opportunities with positive expected payoff, the Kelly system usually calls for a diversified investment strategy that yields overall profits even when some of the investments don't pay off.

Chapter 9 🎲 🎲

Zero-Sum Games: The Abstraction of Hardball

"Generally speaking, the way of the warrior is resolute acceptance of death."
MIYAMOTO MUSASHI, *A BOOK OF FIVE RINGS*

*I*n the seventeenth century the great Japanese samurai Miyamoto Musashi wrote a treatise on strategies for the martial arts, *Go Rin No Sho (A Book of Five Rings)*. Although Musashi devoted his life to fighting, he asserted that the way of strategy could be applied to all walks of life. He viewed life as a series of competitive interactions. "The spirit of my school," said Musashi, "is to win through the wisdom of strategy, paying no attention to trifles." This was the same position taken by some of Musashi's contemporaries, the French mathematicians who were studying gambling games. In the early twentieth century the prolific mathematician John von Neumann developed an abstract theory of games that included these disparate disciplines under one roof.

Von Neumann observed that many military and economic interactions had features similar to parlor games. For one thing, such interactions appeared to be strictly competitive. Von Neumann also noticed, as Musashi had noticed three centuries earlier, that strategy was a fundamental concept.

What Is a Game?

A game is an interaction between two or more players, consisting of rules of play, which include a definition of allowable moves, a method for determining when the game ends, and payoffs (either real or symbolic) to the players when the game ends. The players can be people, teams, casinos, companies, countries, alliances, et cetera. We will restrict our attention to two-player games: chess, for example, or NATO versus Yugoslavia.

Payoff Function

A payoff function is a rule that determines the players' payoffs when the game is over, for every possible result of play. The payoff function can be simple, for example, the loser pays the winner a dollar. Or it can be complex. For example, it can be based on the sum of rewards and losses accumulated during the course of play. Players who have different goals for a game may have different notions of what the payoffs should be. For example, we saw that bold play is optimal for a roulette player whose goal is to obtain a specific amount of money without regard for anything else, whereas timid play is optimal for a gambler whose goal is to play for as long as possible before going broke. For games like chess, tennis, or football, the goal is to win.

In more complicated settings, like war, payoffs may not be well-defined, making the comparison of strategies problematic. For example, which military strategy is better, bombing a country and killing thousands of civilians or mounting an invasion, minimizing civilian casualties but incurring many casualties among your own troops? Or is it better to rely on diplomacy to achieve one's goal?

Strategies

A strategy is a rule that specifies a player's move for every possible situation arising during the course of play. Strategies are ranked according to some method of optimizing a player's payoffs as a result of play. A strategy is optimal if it is better than every other strategy relative to specific payoffs. Earlier, we used expected payoff to rank gambling strategies. As we have seen, optimality doesn't mean profitability. For example, all roulette strategies eventually lead to bankruptcy.

Zero-Sum Games

> "In single combat, if the enemy is less skillful than ourselves, if his rhythm is disorganized, or if he has fallen into evasive or retreating attitudes, we must crush him straightaway, with no concern for his presence and without allowing him space for breath. It is essential to crush him all at once. The primary thing is not to let him recover his position, not even a little."
>
> Miyamoto Musashi, *A Book of Five Rings*

Musashi played hardball, the modern term for ruthless behavior. Hardball was also part of von Neumann's game structure. Such games are called zero-sum: One player's loss is the other's gain. Roulette, chess, and football are zero-sum games. Military and economic interactions have traditionally been modeled by zero-sum games. In recent years cooperation has been used in some game models. Later we will discuss cooperation, but for now we play hardball.

The Zero-Sum Payoff

The result of a zero-sum game is a payoff from one player to another. It's convenient to think of the payoff as money and the person receiving the payoff as the winner, even if no money actually changes hands. For example, in a friendly game of tic-tac-toe, we could say that the loser symbolically pays the winner $1, with no money changing hands in the case of a draw. The term *zero-sum* refers to the fact that, since the loser pays the winner, the sum of the payoffs to both players equals zero.

Zero-sum games model ruthless behavior, but since payoffs depend on the results of play, the payoff structure can be anti-Machiavellian. In other words, the means (player's strategy) can affect the end (result of play and corresponding payoff function). For example, the use of nuclear weapons may be the most likely way to win a war, yet, fortunately for us living creatures, the generally accepted immorality of killing a multitude of civilians (along with the threat of retaliation) prevents nuclear nations from using their most powerful weapons. This could be reflected in an appropriate payoff function or could simply be against the rules of play. Unfortunately, warring nations don't always follow the rules of play.

A Simple Win-Lose Game

Generals A and B are retired warriors, too cultured, or just too old, for physical battle. Instead, they play a simple game that might cost them their money but not their lives. Here are the rules of play and resulting payoffs: A bag contains 12 cookies. The generals alternately remove 1, 2, or 3 cookies from the bag. Whoever removes the last cookie is the winner, who then gets paid $1 by the loser. The goal of this game is simple: to win. To put it another way, the goal is to maximize your payoff. There is no middle ground; either you win or you lose.

The generals take turns going first. They experiment with different strategies. General B soon discovers the optimal strategy: Always let your opponent move first. Then remove what's left over between the number of cookies your opponent removes and 4: If your opponent removes 1, you remove 3; if your opponent removes 2, you remove 2; if your opponent removes 3, you remove 1. Since the game starts with 12 cookies in the bag, and since 12 is divisible by 4, by letting your opponent go first, you can finish removing batches of 4 until you finally get the last cookie.

General A soon figures out what's happening and demands that General B go first. General B refuses. Not wanting to be left with nothing to do, the generals agree to modify the game.

This time the generals put 31 cookies in the bag and allow for the removal of 1, 2, 3, 4, or 5 cookies on each turn. Unfortunately, the proverbial cat, or in this case the cookie, is out of the bag, and both generals quickly realize the new, optimal strategy: Go first and remove 1 cookie. Then always remove the difference between the number of cookies your opponent removes and 6: If your opponent removes 1, you remove 5; if your opponent removes 2, you remove 4, and so on. Here's why this is optimal: After you go first and remove 1 cookie, there are 30 cookies left in the bag. Then, since 30 is divisible by 6, you can always finish removing batches of 6 until you get the last cookie.

These guys didn't become generals because they were stupid. They now see the general pattern, so to speak. Suppose there are some number of cookies in the bag, say N, and that the maximum number that can be removed on one player's turn is R (in the previous game, $N = 31$, $R = 5$). If $R + 1$ is a divisor of N, have your opponent move first. If $R + 1$ isn't a divisor of N, you go first and remove just enough cookies so that $R + 1$ is a divisor of the remaining cookies. In either case, your opponent is left with a bag of cookies that is a multiple of $R + 1$. Now, whatever number of cookies your opponent removes, you remove the difference between that number and $R + 1$, so that, in the same way as before, you keep removing batches of $R + 1$ cookies, until you get the last cookie in the bag.

The generals were happy to learn the structure of the game, but their success made the game boring. It was time to eat the cookies and find another game.

Drawing cookies from the bag is called a game of *perfect information*, because the players take turns moving and because each player knows the entire history of the game before making a move. Tic-tac-toe, checkers, and chess are also games of perfect information. War is a game of imperfect information.

A Ritual War Game

In remembrance of things past, and because they ate all the cookies from their previous game, the generals invent a ritual war game. General A will command the powerful, attacking army. General B will lead the weaker, defense forces. Here are the rules of play: General A has three permissible actions: invade, bomb, or use diplomacy. General B has two actions: surrender or resist. This is a game of imperfect information. The generals simultaneously choose a strategy, without revealing their choice in advance. When the strategies are revealed, the game ends and an appropriate payoff is made. The generals, who are making up the rules of this game on the fly, must now determine what *appropriate* means.

General A's goal is to have General B surrender with minimal effort and with as few casualties as possible for the attacking army. General B's goal is to resist and hope that General A will go away. These goals are vague and must be quantified by payoffs before the generals can analyze their possible strategies. Although there aren't unique payoffs that characterize their goals, the generals make a list of game dynamics that provide a natural ranking that any set of payoffs should conform to:

If General B resists and General A chooses diplomacy, nothing much will happen, so the diplomacy payoff should be the best of General B's payoffs when he uses the resistance strategy.

Still going with the resistance strategy, bombing should be preferable to General B over invasion, because an invasion will probably result in General B's surrender, whereas it's easier to resist bombing.

On the other hand, if General B decides to surrender, General A's best choice is diplomacy, since he would be accomplishing his goal with no casualties or other costs of war. Still considering his cost and casualties, if General B will surrender, General A would prefer bombing to invasion.

The generals come up with a set of payoffs that satisfy these conditions and display the resulting game as a *payoff matrix*, which gives a payoff for each strategy pair, represented in an array. In this representation, General A is the row player: His strategies or possible actions are represented by the rows of the payoff matrix. General B is the column player: His strategies are represented by the columns of the payoff matrix. By convention, the entries in a payoff matrix are the column player's payoffs to the row player for the corresponding strategy pairs:

		General B	
		surrender	resist
	invade	6	5
General A	bomb	7	4
	diplomacy	9	-10

Suppose General A chooses to invade. Then, if General B chooses to surrender, he pays General A $6, and if he chooses to resist, he only pays General A $5 (retention of pride, additional cost to General A). Suppose General A chooses to bomb. Then, if General B chooses to surrender, he pays General A $7 (not only does General B surrender, he incurs casualties), and if he chooses to resist, he pays General B $4 (General B can resist bombing, at a cost to General A). Suppose General A chooses diplomacy. Then, if General B surrenders, he must pay General A $9, because General A will achieve his objectives with minimal cost and no casualties, whereas, if General B resists, it is a great loss for General A (the negative payoff means General A pays General B $10), since General B will have outsmarted him.

The Minimax Criterion: Optimal Strategies in a Hostile World

Zero-sum games are purely competitive: One player's gain is the other's loss. In this environment it's prudent to assume that your opponent is an intelligent and merciless adversary. This suggests a procedure: Try for the best possible payoff, assuming the worst from your opponent. This is called the *minimax* criterion and provides a basis for optimality that makes sense in a zero-sum environment. In a zero-sum world, where cooperation isn't permissible, you should always assume the worst from your opponent.

In the game between Generals A and B, since the entries in the payoff matrix are General B's payoffs to General A, General B would like them to be as small as possible. General B knows that General A has the opposite goal and wants the payoffs to be as large as possible. If General B chooses to surrender, his largest possible payoff to General A is $9. If he chooses to resist, his largest possible payoff to General A is

$5. He can thus minimize his maximum payoff to General A by resisting and guaranteeing a payoff of no more than $5. This is General B's minimax strategy; it minimizes his maximum payoff to General A.

On the other hand, General A wants the payoff to be as large as possible. If he decides to invade, his smallest possible payoff from General B is $5. If he decides to bomb, his smallest possible payoff from General B is $4, and if he decides to use diplomacy, his smallest possible payoff to General B is –$10. General A maximizes his minimum payoff from General B by choosing to invade, guaranteeing a payoff of at least $5. This is General A's *maximin* strategy; it maximizes his minimum payoff from General B.

Equilibrium Points: The Value of the Game

General B's minimax strategy guarantees that he will pay General A no more than $5. General A's maximin strategy guarantees that General B will pay him at least $5. Since these strategies for both players achieve the identical result, with minimax = maximin, we say that the game has a "value" of $5, and that the resulting strategies are optimal: General A should invade; General B should resist.

In this game the optimal strategies are called "pure" strategies because there is no randomness. In this case the value is an entry in the payoff matrix, called an equilibrium point. An equilibrium point is the smallest number in its row and the largest number in its column. Matrix games with equilibrium points are easily solved. Any row (or column) containing an equilibrium point is an optimal pure strategy that achieves the value of the game.

Spying

Games with a value have an interesting property: Knowledge that you are playing optimally doesn't help your opponent. Thus, if you are playing optimally, your opponent has no need to spy on you. For example, if General B tells General A that he will be resisting, General A will still invade and still get a $5 payoff, the value of the game. If General A tells General B that he is invading, it won't change General B's optimal resistance strategy. If you don't play optimally, however, your opponent may be able to take advantage of you. For example, if General B decides to surrender, a nonoptimal strategy, and if General A finds out about it, General A can change his strategy to diplomacy and get the $9 payoff. Thus, the only time spying makes sense in a game like this is if you are not playing optimally.

Since the conflict between General A and General B has a value, and since both players know the optimal strategies, General B may as well pay General A the $5 and call it a day. This brings up an important point: Even if you are cunning enough to play optimally, the game might be unfavorable to you. In this case, you can take solace in the fact that if you weren't playing optimally, you could be doing even worse.

The Sure-Thing Principle

The generals' war game not only has a value, it provides an example of the sure-thing principle, formulated by statistician L. J. Savage, which can be informally stated as follows: *If you have a choice of two strategies, X and Y, and X is always better than Y, no matter what the opponent does, you should always select X.*

In the conflict between General A and General B, if General B chooses to resist, he will get a better payoff than if he chooses to surrender, regardless of General A's choice: Paying $5 to General A for resisting is better than paying $6 for surrendering when General A chooses to invade; paying $4 is better than paying $7 when General A chooses to bomb; and getting paid $10 is better than paying $9 when General A chooses diplomacy. Thus, no matter what General A chooses, General B should always resist. For this particular game, resistance *dominates* surrender. This is the strongest form of optimality. Of course, if the payoffs are changed, it can be a different story.

More about Payoffs

"As one judge said to another, 'Be just, and if you can't be just, be arbitrary.'"

William Burroughs, *Naked Lunch*

In a win-lose parlor game like chess, checkers, and the generals' cookie game, having the loser pay the winner $1 is a natural way to quantify the result. In the ritual war game, the determination of payoffs was based on the generals' notion of sensible game dynamics. Someone who disagrees with these dynamics might suggest different payoffs that change the nature of the game. For example, if you believe that it is better to surrender than to incur any casualties, you would not be happy with the generals' payoff matrix, since it leads to resistance as General B's optimal strategy. In general, game payoffs can be completely arbitrary.

Negotiated Settlements

Good generals hate to surrender, but what about a negotiated peace deal in which the opposing sides compromise— what they call in business a win-win situation? Unfortunately, zero-sum games don't allow this sort of thing. The structure of a zero-sum game, in which one player's gain is the other's loss, prohibits the opportunity for a win-win situation.

The Many-Move Game

The generals' ritual war game lasted only one move, but most games, from chess to war, take many moves to complete. A strategy for a many-move game allows the player to make a decision about the current move based on the previous history of play. For example, a chess player's move depends on the current board position, which is the cumulative result of many moves. In a war, decisions are based on what has happened to date. If the enemy has been weakened, a different dynamic exists than if reinforcements have just arrived. The generals' one-move war game is a simplified model that may or may not provide insights into a more complicated situation.

A Game with No Equilibrium Point

Since the generals have solved their war game, they are again left with nothing to do. Rather than face the horrifying prospect of life without war, they invent another game. In this game, General B can attack General A's forces from either the north or the south, while General A can defend against the attack from either direction. If General A defends the correct position, General B pays him $1. If General A defends the wrong position, he pays General B $1. The payoff matrix is as follows:

| | | General B | |
| | | attack | |
		north	south
General A	north	1	−1
defend	south	−1	1

This game has no equilibrium point. In other words, there is no number that is the largest in its row and the smallest in its column. It doesn't matter which row General A picks: His maximin payoff, the most he can guarantee receiving from General B, equals $−1. It also doesn't matter which column General B picks: His minimax payoff, the least he can insure having to pay General A, equals $1. Thus, minimax is not equal to maximin. The game doesn't have a value in the current structure, and there are no optimal pure strategies.

This game is unstable. Either player can win $1 by having advance knowledge of the other's strategy. In repeated play, a studious player might recognize and exploit the opponent's pattern of play. For example, if General B alternates north, south, north, south, sooner or later General A will recognize the pattern and defend the north against B's northern attack and the south against B's southern attack, always winning $1. Sooner or later General B will get wise and change his strategy to exploit A's pattern, and so on, ad nauseam. Unlike games with equilibrium points, in which you always pick the same strategy, here it makes sense to vary your behavior in order to keep your opponent guessing. Musashi put it this way:

> The mountain sea spirit means that it is bad to repeat the same thing several times when fighting the enemy. There may be no help but to do something twice, but do not try it a third time.

If you once make an attack and fail, there is little chance of success if you use the same approach again. If you attempt a technique that you have previously tried unsuccessfully and fail yet again, then you must change your attacking method. If the enemy thinks of the mountains, attack like the sea; and if he thinks of the sea, attack like the mountains.

Randomness to the Rescue

Varying your strategy doesn't help if you leave clues. For example, we have seen that if General B varies his strategy by alternately choosing north, south, north, south, a predictable pattern is established that General A can use to his advantage. The idea is to vary your strategy without leaving a pattern. There is only one way to do this: randomness.

Suppose General B privately tosses a coin before each play. If the coin lands heads, he picks north; if it lands tails, he picks south. Unless General A sees the result of General B's toss before picking his strategy, he has only a 50 percent chance of guessing what B will do. Strategies that use randomness are called *mixed* strategies. Strategies with no randomness, like the optimal strategies in the previous game, are called *pure* strategies. As we saw when we analyzed casino games, an appropriate way to analyze strategies that use randomness is to compute the average, or "expected," payoff. In this case, when at least one player submits a mixed strategy, expected payoff is the average of the payoffs relative to the appropriate probabilities. For example, if General A picks north against General B's coin-tossing strategy, General A's expected payoff = $.5 \times 1 + .5 \times (-1) = 0$. If General A picks south, his expected payoff = $.5 \times (-1) + .5 \times 1 = 0$. In other words, if General B uses the coin-tossing strategy, General A's average payoff equals 0, no matter what General A chooses. This mixed strategy has the same stabilizing effect as an optimal strategy in a game with an equilibrium point.

The coin-tossing strategy is optimal in the following sense: If General B uses any other strategy, mixed or not, General A can exploit it and increase his expected payoff. For example, suppose that, instead of tossing a coin, General B randomly draws a ball from a box containing two tickets marked north and one ticket marked south and then chooses the strategy on the selected ticket. Then, if General A always picks north, his expected payoff = 2/3 × 1 + 1/3 × (–1) = 1/3. In other words, General B's "2/3, 1/3" strategy gives General A an average profit of 33 1/3 cents per game in repeated play, a higher payoff than if General B used the coin-tossing strategy.

Since the generals' optimal strategy requires tossing a coin, *skill* requires *chance*. If you don't randomize, you will be a long-run loser.

Matching Fingers

The generals' current game is a version of the matching-fingers game and can be rephrased as follows: General A and General B simultaneously show either one or two fingers. If they match, General A wins. If they don't, General B wins. Those of us who played this game when we were kids learned the importance of mixed strategies. Most people don't actually toss coins or draw balls from boxes but instead pick at random mentally. It has been shown that the random-number generators in our minds are inadequate. For example, people trying to write down sequences of mental coin tosses tend to alternate heads and tails more often and generate fewer streaks of all heads or all tails than actual coin tossing (an interesting discussion of this can be found in Peterson). The bottom line: A cunning opponent can exploit shoddy randomness. In a game for serious money, if you don't use the proper equipment (coins, dice, balls in boxes, et cetera), you may end up a loser.

A Revised Game

It is a war of attrition. In 10,000 plays, General A wins 5,047 times and General B wins 4,953 times, close to a 50-50 split and yielding General A a profit of $94. Since the payoffs are based entirely on credit, it is a pointless ritual. The war-weary generals forget why they started playing this game in the first place. Even though the thrill is gone, it is difficult to end the game and move on to something else: When General A suggests quitting, General B accuses him of not giving him a chance to win his $94 back. Later, when General B suggests

quitting, General A calls him a coward. It seems as though
the game will continue until one of them dies. Then one day,
General A has an inspiration: to arbitrarily alter the payoffs,
as follows:

		General B	
		attack	
		north	south
General A	north	−20	30
defend	south	5	−25

The pristine beauty of the new game, whose arbitrary
payoffs have nothing whatsoever to do with reality, excites
the generals. When they start playing, each general uses the
coin-tossing strategy from the previous game, making each of
the four payoffs equally likely and yielding the following
expected payoff (from B to A):

$$1/4 \times (30 - 20 + 5 - 25) = -\$2.50$$

The coin-tossing strategy is profitable for General B, netting him an average payoff of $2.50 per play. After a while,
General B decides to change his strategy. He realizes that if
he always picks north while General A keeps tossing a coin,
he will win $20 about half the time and lose $5 about half the
time, for an average profit of $7.50 per play. This is better
than winning an average of $2.50 per play, so General B
switches strategies. This works for a while, but as soon as
General A notices what is happening, he starts picking south
every time, yielding him a profit of $5 per play. As soon as
General B realizes that General A has changed strategies, he
starts picking south every time, yielding him a profit of $25
per play. General A catches on fast and counters by always
picking north. The battle is on.

After considerable trial and error, with periodic time-outs
for pizza, General B settles on the following mixed strategy,
which he calls the 11-5 box strategy: Put 11 balls marked
north and 5 balls marked south in a box. Draw a ball at random from the box, and act accordingly.

Proceeding with this strategy, General B picks north with probability 11/16 and south with probability 5/16. The interesting result: No matter what strategy General A uses, General B's profits average $4.38 per play!

Here's how it works: If General A picks north, he loses $20 to General B an average of 11 times in 16 plays and wins $30 an average of 5 times in 16 plays. Expected payoff: 11/16 × (–20) + 5/16 × 30 = –$4.38.

If General A picks south, he wins $5 from General B an average of 11 times in 16 plays and loses $25 an average of 5 times in 16 plays. Expected payoff: 11/16 × 5 + 5/16 × (–25) = –$4.38.

Either way, General B makes an average profit of $4.38 per play. Since this is true for either choice of General A, it must happen for any mixed strategy as well. In other words, General B's 11-5 box strategy is an optimal mixed strategy and creates a state of equilibrium in the same way as the coin-tossing strategy of the previous game and the optimal pure strategies of earlier games.

Whenever General B uses this strategy he guarantees an average profit of $4.38, regardless of what General A does. In addition, if General B uses any other strategy, General A has a strategy that will lower General B's $4.38 average profit.

It turns out that General A has his own optimal strategy: the 3-5 box strategy. If General A puts 3 balls marked north and 5 balls marked south into a box, draws a ball at random, and picks the chosen strategy, his average payoff to General B will be $4.38, no matter what General B does. If General A uses any other strategy, he could be exploited by General B and do worse.

Tanya the Trader Uses Game Theory to Guarantee a Profit

The generals' broker, Tanya the Trader, has $100,000 to invest. She has discovered two exciting investment opportunities: Intergalactic Holding Corp. (IHC) and Mega Ltd. (ML). ML performs best during a recession, and IHC performs best during inflation. Tanya estimates annual percentage profits and puts this information into a payoff matrix:

		Economy	
		Inflation	Recession
Tanya	IHC	25	−5
	ML	−10	40

The players of this game are Tanya and the economy. If Tanya invests in IHC and there is an inflationary economy, she makes a 25 percent profit; however, if there is a recession, she incurs a 5 percent loss. If Tanya invests in ML and there is inflation, she incurs a 10 percent loss; however, if there is a recession, she makes a 40 percent profit. The payoff matrix makes this decision look like a zero-sum game; however, Tanya would be considered odd if she believed that the economy was a clever opponent who wanted to defeat her.

Although she has a bumper sticker that says JUST BECAUSE I'M PARANOID DOESN'T MEAN THAT NOBODY'S OUT TO GET ME, Tanya doesn't believe that the economy is out to beat her in this game. Instead, she makes clever use of the zero-sum structure. After careful deliberation, Tanya determines that her optimal mixed strategy for this game is the 5-3 box strategy: Put 5 balls marked IHC and 3 balls marked ML into a box. Draw a ball at random, and use the appropriate strategy.

When Tanya uses the 5-3 box strategy she is guaranteed an average profit of V = 11.875 percent (the value of the game). Any other strategy could result in a lower profit if the economy takes a turn for the worse. The 11.875 percent average profit applies when Tanya plays the game repeatedly under the same conditions. Unfortunately, the stock market is a dynamic process and conditions change. Also, Tanya's clients, the generals, want short-term guarantees. Interestingly, there's another way to use this game's optimal strategy.

Diversification

Instead of using the 5-3 box strategy to randomly select an investment, Tanya uses the same strategy, which we now call the 5-3 diversification strategy, to find the fraction of her money to invest in each stock: Invest 5/8 of her bankroll ($62,500) in IHC and 3/8 ($37,500) in ML.

In this way, by using the same strategy but with a different meaning, Tanya guarantees a profit of 11.875 percent, regardless of the state of the economy. In Tanya's model, her randomness and risk disappear with diversification. (In real life things are fuzzier, because profit estimates and "states of the economy" are imprecise.) Although other investment allocations could do better if Tanya got lucky, they could also do worse. Using the optimal mixed strategy in this way is an interesting application of game theory and yields the highest profit Tanya can guarantee.

Note: This diversification strategy only makes sense if the investment game has a positive value. Otherwise, playing optimally in this way will yield a guaranteed loss.

Another note: There are other methods of diversification, such as the Kelly system, mentioned in Chapter 8, that can yield better long-term results, but they depend on accurate estimates of probabilities for future states of the economy.

Optimal Mixed Strategies: The Minimax Theorem

The main mathematical result of game theory, first proved by John von Neumann, is called the minimax theorem, which asserts that every zero-sum matrix game has a value, V, along with optimal strategies, possibly mixed, for each player, as follows:

1. If the payoff matrix has an equilibrium point, it equals the value of the game. In this case, an optimal pure strategy (no randomness) consists of picking either a row or a column that contains the equilibrium point. The row player using an optimal pure strategy guarantees a payoff of at least V, no matter what the column player does, and the column player using an optimal pure strategy guarantees a payoff of at most V, no matter what the row player does.

2. If the payoff matrix has no equilibrium point, there are optimal mixed strategies for each player (bring in the balls and boxes). Using an optimal mixed strategy, the row player guarantees an expected payoff of at least V,

no matter what the column player does, and the column player guarantees an expected payoff to the row player of at most V, no matter what the row player does.

In both situations, knowledge that a player is using an optimal strategy doesn't help the opponent. In addition, a knowledgeable opponent can exploit a player who is not playing optimally. When the payoff matrix doesn't have an equilibrium point and randomness is introduced in the form of mixed strategies, individual results vary. Only average results in repeated play are assured.

The minimax theorem holds for any zero-sum matrix game, even if there are a billion entries in the payoff matrix. In fact, all games with a finite number of possible moves fall into this category, even if they do not neatly fit into matrix form. Unfortunately, finding an optimal strategy for a complicated game can be a difficult task, even if you know that one exists.

Optimal mixed strategies guarantee optimal average payoffs, but a player's payoff may suffer if the opponent discovers the result of the coin toss or random drawing before making a move. In military and economic encounters, where even the rules of play are vague, great lengths are taken to find out what the opponents are doing. Espionage will be around wherever there are zero-sum games of imperfect information with no obvious optimal strategies.

Sporting events provide an example of mixed strategies. One of the secrets of sports success is catching the opponent off guard. Regardless of athletic ability, players and teams that become predictable soon hit the skids.

Optimal mixed strategies achieve predicted results in repeated play. In fact, most gamelike interactions involve repeat encounters. Companies and countries, even bitter rivals, develop repetitive routines. Don't fool yourself into thinking that you will only play a game once.

Optimal Strategies for Games of Perfect Information

The minimax theorem guarantees that matrix games of imperfect information have a value, along with optimal, possibly mixed strategies. What about games of perfect information, like chess and checkers? These games have a value, along with optimal, pure strategies. (There is no reason to use randomness in a game of perfect information, because your opponent knows the result of your move before having to move.)

For example, it can be shown theoretically that there exists an optimal chess strategy that will either always guarantee one player (white or black pieces) a victory, no matter what the opponent does, or always guarantee each player a draw. The same is true for all other board games with perfect information.

Chess is a complex game. Nobody knows the optimal strategy, only that it exists. In recent years, chess-playing computer programs have been able to consistently beat chess experts. As computer technology progresses, chess-playing programs will get faster and be able to quickly analyze many layers of move possibilities, eventually leaving human chess players in the dust.

A Game Without a Value

Generals A and B get bored again and, necessity being the mother of invention, devise a new game. Each player simultaneously picks a positive integer. Whoever picks the larger number is the winner and gets $1 from the loser. If both players pick the same number, the game is a draw, and no money changes hands.

In this game, each player has an infinite number of choices, and so the game cannot be put into matrix form. In the process of playing this game, General A says to General B: "Anything you can pick, I can pick bigger." General B's response is, "Anything you can do, I can do too." In other words, whatever number one player picks, the other can pick a larger one. Nobody can guarantee anything, even with mixed strategies. This game has no value and no optimal strategies. It is unstable, not just because each player has an infinite number of choices but because the choices are unbounded, that is, there is no largest number. Trying to win a game like this can get military strategists into a situation commonly known as an arms race. Fortunately, there are only a finite number of resources for building weapons, and so arms races are, in a sense, bounded.

Infinite Games

Even though there may be only a finite number of possible moves, a game may never end. Some military conflicts continue from generation to generation and may even take place over centuries. Economic encounters between countries may also persist for extended periods. In such situations there may be rules of play but no final payoff, only an ongoing ritualistic tally of some sort.

Chapter 10 🎲🎲

Prisoner's Dilemma: A Game with Cooperation

"I never hold a grudge. As soon as I get even with the son of a bitch, I forget about it."
    W. C. FIELDS

*I*n zero-sum games there is no cooperation. They are encounters of pure competition, in which one player's gain is the other's loss and you play assuming that your opponent is out to defeat you. To play optimally, you may need a mixed strategy, in which chance makes your opponent uncertain of your move. Nasty behavior, or, to put it politely, trying to defeat your opponent, is part of the zero-sum structure. This is terrific for tennis, football, and chess, and it is sometimes swell for military, political, and economic encounters, but interactions that appear competitive may also allow cooperation. Even if cooperation becomes part of the game, nastiness may be prudent in order to avoid being exploited. Those who are nasty when cooperation is possible may miss out on long-term opportunities. Unfortunately, cooperation is impossible unless the zero-sum paradigm is replaced by a game structure that will allow both cooperation and competition.

A Prisoner's Dilemma Primer

For interactions that can involve both cooperation and competition, a basic question arises: *How can you cooperate yet avoid being exploited?*

Such interactions can often be represented by a simple game called Prisoner's Dilemma, formulated in the 1950s by mathematician A. W. Tucker. In recent years researchers have used this game to model interactions in such diverse fields as biology, economics, psychology, political science, and war.

In its original form Prisoner's Dilemma involved two prisoners accused of committing a crime together who are put in isolated cells and encouraged to confess and rat on each other. The evidence is scant: If both remain silent, they both receive a lighter sentence than if both confess; however, if one confesses and rats on the other, while the other remains silent, the one who confesses goes free, and the one who remains silent gets a stiff sentence. Here is the dilemma: If both prisoners remain silent (unknowingly cooperate with each other), they both do better than if they both confess (unknowingly being uncooperative), yet if one is silent and the other confesses, the one who is silent gets double-crossed and sent to jail, while the one who confesses goes free. Since the prisoners are in isolated cells, this is a game of imperfect information.

Here cooperation means being silent and not ratting on one's partner in crime, however, doing so may result in the double cross and jail time.

The Generals' Version

Generals A and B are tired of their warring ways and decide to play a game that allows cooperation. Each player will have two choices: to cooperate or to be nasty. (*Nasty* is an arbitrary term that means noncooperative.) This is a game of imperfect information: the generals move simultaneously and without knowledge of their opponent's move. If both cooperate, both do better than if both are nasty; but if one general is nasty and the other is cooperative, the nasty general exploits the cooperator and gets a high payoff. Their dilemma: How can they cooperate yet avoid being exploited?

Since the generals can cooperate, it follows that one general's gain isn't necessarily the other's loss. Thus, entries in the payoff matrix can't be single numbers that represent the column player's payoff to the row player for a given pair of strategies. Instead, they are pairs of numbers that represent the payoff to each player from an outside source, say, nature, or a master of ceremonies. In the following payoff matrix, the left number represents the row player's (General A's) symbolic payoff (say, in dollars) and the right number represents the column player's (General B's) symbolic payoff:

		General B	
		Cooperate	Be nasty
General A	Cooperate	(3, 3)	(0, 5)
	Be nasty	(5, 0)	(1, 1)

Payoff Explanation

3 = Reward payoff for mutual cooperation
1 = Punishment payoff for mutual nastiness
5 = Temptation payoff for taking advantage of a
cooperative opponent
0 = Sucker payoff for being cooperative with a
nasty opponent

If both generals cooperate, they each get $3, the reward payoff. If both are nasty, they each get $1, the punishment payoff. If one general cooperates and the other is nasty, the nasty general gets $5, the temptation payoff, and the cooperator gets 0, the sucker payoff.

Prisoner's Dilemma payoffs are arbitrary, subject to the following two conditions:

1. Temptation > reward > punishment > sucker.

2. Reward is better than the average of sucker and temptation: Reward > (sucker + temptation)/2.

Condition 1 establishes the order of the various interactions between the players. Condition 2 guarantees that the players can't take turns being nasty and cooperative and do better, on the average, than mutual cooperation. Any set of payoffs that satisfies these conditions is Prisoner's Dilemma.

The generals start by playing Prisoner's Dilemma once. General A immediately notices that being nasty dominates cooperation: No matter what the opponent does, it's always better to be nasty than to cooperate: If General B cooperates, General A gets $5 for being nasty, $3 for cooperating. If General B is nasty, General A gets $1 for being nasty, 0 for cooperating. The sure-thing principle applies: General A should be nasty. Since the game is symmetric, General B should also be nasty. Again, we encounter the dilemma: Although being nasty gives a higher payoff than cooperation regardless of the opponent's choice, if both generals cooperate, both do better than if both are nasty.

Familiarity can breed cooperation, not contempt. If you only play once, it follows from the sure-thing principle that being nasty is optimal. If you're never going to see your opponent again, there's no strategic reason to be nice. This sounds dismal, but most gamelike interactions, whether between companies, countries, or couples, are repetitive. Even animals roaming the jungle have well-established foraging patterns, causing repeat encounters with other animals. (As one hyena said to another, "I think I recognize that lion.") If the game lasts more than one move, there may be cause for cooperation.

The generals, whose existence is based on repetitive, game-playing activities, devise a new version of the game: Play twice. Their payoff will be the sum of their payoffs from each move. Now their strategies can be more complicated, because they can base their second move on the opponent's history of play (the first move). General B decides to use the following strategy for the two-move game: Cooperate on the first move. If General A cooperates on the first move, cooperate on the second move, but if General A is nasty on the first move, be nasty on the second move.

General A decides to be nasty on both moves. He gets the $5 temptation payoff on the first move and the $1 punishment payoff on the second move, for a total of $6. After they play and discuss their strategies, General A realizes that if he had cooperated on the first move and had been nasty on the second move, he would have received the $3 reward payoff on the first move and the $5 temptation payoff on the second move for a total of $8. Thus, against this strategy, always being nasty is not optimal, as it is in the one-move game. The sure-thing principle no longer applies. It may make sense for the generals to cooperate in repeated play.

Biologists have shown that when the end is near, creatures that have previously cooperated may become nasty. Bacteria that have lived harmlessly within an organism may attack the organism if it becomes unhealthy. Animals who live in harmony under ordinary circumstances may get nasty under adverse conditions. Predators look for unhealthy prey. To paraphrase the slogan on the Statue of Liberty: "Give me your tired, your sick, and your wounded, and I'll eat them." In business, a company having financial problems may suddenly find that former partners are no longer friendly. The same goes for politics. The fact that nobody wants to support a losing cause is starkly illustrated by the dynamics of Prisoner's Dilemma: Since there is no future beyond the last move, the last move is like a lone encounter, and you should be nasty.

The generals knew that since nastiness is optimal in the one-move game, the only potential for cooperation is in repeated play. They realized that since repeated play ends on the last move, the last move is equivalent to the one-move game, so you should be nasty on the last move. This leads General B to a slippery slope of reasoning: If you're nasty on the last move, the future effectively ends on the second-to-last move, so it makes no sense to cooperate then either. If it makes no sense to cooperate on the second-to-last move, it makes no sense to cooperate on the third-to-last move, and so on, down to the first move. General B concludes that cooperation collapses like a house of cards. General A sees a way out: In most real-life situations, even though the game eventually ends, there is no predetermined last move. The same structure can be used in Prisoner's Dilemma.

The Iterated Game

The generals decide that the way to set up their Prisoner's Dilemma game so that there is no predetermined last move is to use chance to decide when the game ends. They will toss a coin after each move to decide whether or not to continue play. After a few rounds of doing this, they realize that this system doesn't allow much flexibility: After each move there is always a 50 percent chance that the game will end, so the game will always last for an average of two moves. They decide to use a more flexible method of randomly deciding when to stop play: They put some balls marked "stop" and some balls marked "continue" into a box. After each move, they draw a ball at random from the box. If a stop ball is selected, the game stops. If a continue ball is selected, the ball is replaced in the box and the game continues. The chance of stopping or continuing depends on how many balls of each type they have put in the box, so that the average length of a game can be whatever the generals want it to be. A general's total payoff for this version of the game is the sum of payoffs for all the moves until the stop ball is selected, and the game ends. The original payoff matrix represents one move of the bigger game. This form of the game is called *Iterated Prisoner's Dilemma*.

Here's an example of the iterated game. The generals put three continue balls and one stop ball in the box, so that after each move the chance equals 1/4 that the game will stop. When they first play this game, the generals both cooperate on each move, so their payoff on each move is always $3. By chance, the stop ball is selected after the third move, so each general gets a $9 payoff. If both generals had always been nasty, the total payoff to each would have been $3. If General A had always been nasty and General B had always cooperated, the total payoff to General A would have been $15, and the total payoff to General B would have been 0.

When discussing gambling games, we used the law of averages to assess long-run results. From the casino's perspective, the long run means lots of action: One gambler making many bets is the same as many gamblers making one bet each. In Prisoner's Dilemma the situation is different. Playing against one opponent for many moves is different from playing many one-move games, because you can learn from your opponent's history of play.

Rethinking Old Concepts

To win or not to win—that is the question. After playing Iterated Prisoner's Dilemma a number of times, General A has an epiphany: In Prisoner's Dilemma, unlike zero-sum games, one player's gain is not necessarily the other's loss, so there is no reason to assume that your opponent is out to defeat you. Thus, in Prisoner's Dilemma, the meaning of winning is transformed: As long as you do well, it doesn't matter if your opponent does better. This, General A realizes, is a paradigm shift for games. In the new setting, winning means only that you maximize your total payoff. If this means that your opponent, in this case General B, does better than you, that's fine. Both you and your opponent can succeed. Or you both can fail. When General A tells General B about his great insight, General B accuses him of getting hinky because he has been away from battle for so many years.

Prisoner's Dilemma is a mathematical metaphor for real-life situations. Here *being nasty* has meaning only in the context of a particular encounter. For example, to a couple having a dispute, it could mean having a verbal disagreement. In a business negotiation, it might mean not making a deal. In a dispute between countries, it could mean starting a war. Mahatma Gandhi, the great Indian spiritual leader and proponent of non-violence, practiced civil disobedience as a form of nasty behavior when his detractors refused to cooperate. The Reverend Martin Luther King, Jr., exercised nonviolent civil disobedience when confronted by racism. For both Gandhi and the Reverend King, refusing to cooperate didn't mean resorting to violence.

Even simple metaphors can be meaningful. Although Iterated Prisoner's Dilemma involves repeated play, each player has only two possible choices for each move. Real-life situations are usually more complicated than this, but insights can be gained from simple games. General A realized this when he had the epiphany that changed the way he thinks about winning and losing. It must be acknowledged, however, that Generals A and B seldom leave their cluttered apartment, and so simple games, which occupy most of their time, have more meaning in their lives than in the lives of most people.

A Strategy Sampler

A strategy for Iterated Prisoner's Dilemma is a rule that specifies whether to cooperate or be nasty on the current move, given the history of play. Here are some of the generals' favorite strategies for the iterated game:

Always cooperate: Cooperate on every move.

Always be nasty: Be nasty on every move.

Retaliate: Cooperate as long as your opponent cooperates, but if your opponent is ever nasty, be nasty forever after.

Tit for tat: Cooperate on the first move. Thereafter, do whatever the opponent did on the previous move: Cooperate if the opponent cooperated. Be nasty if the opponent was nasty.

Tit for two tats: Cooperate on the first move. Thereafter, cooperate unless the opponent has been nasty on the previous two moves, in which case, be nasty on the next move.

Be random: On each move toss a coin. If the coin lands heads, cooperate. If the coin lands tails, be nasty.

Alternate: Cooperate on every even-numbered move, and be nasty on every odd-numbered move.

Fraction: Cooperate on the first move. Thereafter, cooperate if the fraction of times your opponent has cooperated until now is greater than 1/2, and be nasty if the fraction of times your opponent has cooperated is less than or equal to 1/2.

Sophisticated fraction: Cooperate on the first move. Thereafter, move as follows: If the opponent has cooperated C times and has been nasty N times, put C balls marked cooperate and N balls marked nasty in a box, then randomly select a ball, and proceed as indicated. You are then cooperating with probability $C/(C+N)$ and being nasty with probability $N/(C+N)$, mimicking the opponent's aggregate behavior to date.

The Generals Offer a Few Tips about Cooperating

After playing thousands of rounds of Iterated Prisoner's Dilemma, the generals decide to pass on to future generations some of the knowledge they have gained:

If your opponent disregards your behavior, be nasty. If your opponent will always be nasty, regardless of what you do, you should also always be nasty, because nastiness against nastiness yields the punishment payoff of 1, whereas cooperation against nastiness yields the sucker payoff of 0. If your opponent will always cooperate, regardless of what you do, you should again always be nasty, because being nasty when your opponent cooperates yields the temptation payoff of 5. In fact, you should always be nasty against any opponent who doesn't pay attention to what you are doing. In other words, if your opponent's strategy isn't affected by your behavior, you may as well be nasty.

If you want cooperation, forget about the love generation. Since mutual cooperation is better for both players than mutual nastiness, it might seem that by always cooperating, a player will set a good example that the opponent can then follow. Isn't this what the love generation of the 1960s was all about? Unfortunately, always cooperating is foolish game play that allows an opponent to exploit a player on every move without fear of retribution. May the love generation rest in peace.

The threat of retaliation encourages cooperation. Suppose you use the retaliation strategy: Cooperate for as long as your opponent cooperates, but if your opponent is ever nasty, be nasty forever after. Suppose the iterated game lasts ten moves. An opponent who is nasty on every move will get the 5-point temptation payoff for the first move (you cooperate) and the 1-point punishment payoff for the remaining nine moves (you're always nasty in retaliation to your opponent's initial nastiness), for a total payoff of 14 points. On the other hand, an opponent who cooperates on every move allows you to cooperate as well, netting you both a reward payoff of 3 points per move, for a total of 30 points. In a many-move game against retaliation it is better to always cooperate than to be nasty. Unfortunately, retaliation is unforgiving. Even though it encourages cooperation, if you are nasty just once, retaliation will never cooperate again.

Tit for tat encourages cooperation. Suppose you use tit for tat. Start by cooperating, and then always do what your opponent did on the previous move: When your opponent cooperates, you cooperate, and when your opponent is nasty, you are nasty. The strategy is simple. If your opponent cooperates on every move, so will you, and you will both always get the 3-point reward payoff. If your opponent is nasty, you will return the nastiness on the next move, but as soon as your opponent cooperates, you will cooperate again. If a nasty opponent doesn't get the message, you both suffer. If a nasty opponent has a change of heart, you both benefit.

Always cooperate against tit for tat. If the game lasts for a while, it is optimal to always cooperate against tit for tat. Any other strategy will do worse. This is a new way to analyze a game: Instead of finding strategies that do well against every opponent, use tit for tat to force an opponent to cooperate in order to do well. This is a Prisoner's Dilemma version of an optimal strategy. In this context, for cooperation to work, the game must last for a while. In other words, the ratio of continue balls to stop balls in the box must be large.

General A says, "Remember the paradigm shift." In the traditional, zero-sum setting, winning, or getting a good score, goes along with defeating, or doing better than your opponent. In Iterated Prisoner's Dilemma, playing to get a better score than your opponent can actually cause a decrease in your own score, whereas getting a high score might necessitate allowing your opponent to get an even higher score. For example, if you are always nasty against tit for tat, even though you will beat tit for tat (get a higher score), your score will be lower than if you had cooperated.

General B tells a story about tit for tat. In an ancient land, there was a wise man who dressed like a beggar. Once, after a desert trek, the wise man went to a bathhouse. The arrogant bathhouse attendant was disgusted by the beggar and led him to a moldy basement shower. After the traveler showered, the attendant gave him a dirty towel. The wise man thanked the attendant profusely and gave him a large gold coin. The attendant was astonished that a beggar would have such a coin, let alone give it to him as a tip.

The next week the wise man returned to the bathhouse. This time the attendant treated him like royalty, escorting him to a large, sunny shower stall, giving him a stack of clean towels, powder, the works. Afterward, the wise man thanked the attendant and gave him a penny.

"What's this?" the enraged attendant barked. "Last week, when I treated you like a beggar, you gave me a gold coin. Now, when I treat you like royalty, you give me a penny!"

The wise man replied, "The gold coin was for today. The penny is for last week."

The Generals Do Some Research

There are numerous "intuitive" strategies for Iterated Prisoner's Dilemma, many of them more complicated than tit for tat. In order to learn more about Prisoner's Dilemma strategizing, the generals did some research. Here is what they discovered.

In the early 1980s political scientist Robert Axelrod held two Prisoner's Dilemma tournaments to determine which strategies did well in actual play. Strategies were submitted as computer programs and were pitted against one another in a round-robin tournament, so that each strategy played every other strategy. The winner was the strategy with the highest total score.

Fifteen strategies were in the first tournament. Each match-up consisted of five games of approximately 200 moves. The score for each strategy was its average score for the five games. Half of the strategies in the tournament were nice, that is, never the first to be nasty, and the other half were not nice. All the nice strategies finished ahead of all the not-nice strategies. The winner was tit for tat, submitted by Anatol Rapoport from the University of Toronto.

The second tournament was advertised more widely than the first. There were 62 entries. Every contestant in the second tournament was given the results of the first tournament. Players in the second tournament knew all about tit for tat. Tit for tat won again.

Why Tit for Tat?

Tit for tat is a good strategy because it is cooperative yet can't be exploited. Although tit for tat can never win an individual game in the traditional sense, that is, it can never get a higher score than its opponent, it won both tournaments by getting the highest total score. This is because it did well against nice strategies, getting the large payoff for mutual cooperation, and yet nasty strategies couldn't exploit it.

Axelrod also used Prisoner's Dilemma strategies in an evolutionary model, and found that survival of the fittest—

strategy, that is—ruled the day. A computer program generated a sequence of Prisoner's Dilemma tournaments. After each tournament, strategies were rewarded by having copies of themselves added to the strategy population in proportion to their score. In other words, strategies that did well in a tournament reproduced more than strategies that did poorly.

As cooperative strategies increased in number, they had more interactions with themselves and, because of mutual cooperation, kept growing. In the beginning, nasty strategies did well, but only at the expense of strategies that could be exploited. As time passed, the cooperative strategies that were exploited by the nasty strategies became extinct. Eventually, the nasty strategies had no one left to exploit, so they died too. In other words: *When the prey becomes extinct, the predator dies too.*

After a thousand generations, tit for tat was the top strategy and growing at a faster rate than any other. The nastiest strategies were not the fittest.

Tit for tat did well in Axelrod's computer simulations because it got high scores from cooperation with nice strategies and didn't allow nasty strategies to exploit it. But this doesn't work in a hostile environment: There must be cooperative strategies with whom to cooperate. The message is: If you can't find friendly cooperators, either find a nicer environment or be nasty.

What If You're Not as Strong as Your Opponent?

Prisoner's Dilemma is a symmetric game: Each player has the same strategies and payoffs. On the other hand, there are many interactions with dynamics similar to Prisoner's Dilemma but in which there is a lack of symmetry. If the boss treats a worker unfairly, the worker may not be able to return the nastiness. If a big country invades a small country, the small country may not have the means to reciprocate. If a landlord refuses to fix a tenant's broken heater, the tenant may not be able to do much about it. If a cable television company arbitrarily raises its rates, there's not much a subscriber can do. (One year, when the rates kept rising, the generals threatened to declare war on their cable television provider. Since they no longer commanded armies, they decided to get a satellite dish instead.) If a large factory is spewing pollutants into the air, nearby residents have no easy solution. In situations like these, where there is little recourse to nasty treatment and quitting the game is not feasible, it may be necessary to form alliances. Labor unions, special-interest groups, and alliances of nations have traditionally provided a way to equalize power, in non-symmetric games.

Partisan Politics

Politics in Washington has always been competitive, but somehow Republicans and Democrats have managed to cooperate and compromise to get things done. In the late 1990s things got ugly. Partisan acrimony reached new heights with the impeachment trial of President Clinton, although various forms of nastiness had been swirling around Congress for some time. One of the leaders of the new nastiness was Republican congressman Newt Gingrich, who became Speaker of the House in 1995 but who was forced out by his own party at the end of 1998. In a *San Francisco Examiner* article entitled "GOP's Fine Young Cannibals Devour Gingrich" (November 8, 1998), Christopher Matthews chronicled Gingrich's rise and fall from power. Matthews pointed out that "Gingrich won and held the nation's highest legislative office chewing on the bones of his rivals, championing the raw appetites of his allies." Later in the article Matthews quoted Mark Johnson, former Democratic House Speaker Jim Wright's spokesman, as saying, "Nine and a half years ago, Jim Wright, in his second term as Speaker, stood in the well (of the House) and decried what he called the 'mindless cannibalism' that had forced him to resign. He was referring to the partisanship that had taken over the House that had made everything a zero-sum-game." Matthews commented, "While past leaders could rely on strong personal ties to other lawmakers to stay in power once they got there, Gingrich depended entirely on his own ferociousness rather than the comradeship of his political allies."

The message is clear: In an environment that allows both cooperation and competition, nasty behavior, though it may be successful in the short term, can be self-destructive in the long run.

Collusion Is Cooperation among Companies

In recent years businesses that have been traditionally competitive have learned the value of cooperation. For example, in mid-1999, gasoline prices in California went up about 25 percent. Although this happened for a variety of reasons, and though collusion is illegal, this huge jump at the pump could never have happened if just one company decided to raise prices. Even though there are laws against such things, the government was unable to do much about this questionable type of cooperation among companies.

Collusion sheds an interesting light on non–zero-sum games, in which the payoffs are not from the loser to the winner but are made by some third party. In this case, the third party is the public, which pays the high payoff for mutual cooperation among companies. One sometimes yearns for a world of pure competition, a place where zero-sum rules of the marketplace keep prices at a reasonable level and where collusion is strictly prohibited. General B points to this example as a flaw in General A's epiphany about a paradigm shift.

War

War is perhaps the most vicious and competitive of interactions, yet just as the generals found a cooperative setting for their game playing, researchers have discovered pockets of cooperation on the battlefield. Axelrod and others found that during World War I, troops in the trenches, who faced their enemy counterparts across a battlefield for extended periods, developed tacit rules of cooperation, such as aiming their weapons away from one another in hopes that hostilities would end with a minimum of casualties. Another instance of cooperation in a hostile situation comes when opposing forces negotiate a compromise or peace treaty. Unfortunately, a negotiated settlement is somewhat different from forcing the enemy to surrender. As we shall see in the next chapter, when surrender is demanded by the more powerful side, any hope of cooperation goes up in smoke, and the citizens of the weaker country suffer.

The Generals Call It Quits

After learning how to cooperate without being exploited, the generals decide to stop playing Prisoner's Dilemma and return to zero-sum games. The reason is simple: They can't figure out who is supposed to award them their payoffs.

Chapter 11 ⚅⚅

War Games: NATO Versus Yugoslavia

We live in a world of uncertainty and turbulence. In addition to floods, hurricanes, earthquakes, and other natural disasters, there are disputes, debates, debacles, and other hostile interactions in all corners of the globe. British prime minister Tony Blair has said, "If we wanted to right every wrong that we see in the modern world, then we would do little else than intervene in the affairs of other countries. We would not be able to cope." Yet, on March 24, 1999, NATO started bombing Yugoslavia. Why?

Blair, this time referring to ethnic cleansing in Kosovo, said: "Acts of genocide can never be a purely internal matter." Along the same lines, NATO secretary-general Solana told a news conference during NATO's fiftieth-anniversary celebration, "I think that we are moving into a system of international relations in which human rights, rights of minorities, are much, much more important . . . more important even than sovereignty." Blair, Solana, President Clinton, and the rest of the NATO leaders not only deemed it necessary to intervene in Yugoslavia, they decided to do so by using a strategy of relentless bombing.

The Setup

The Players

The players were the NATO countries on one side and Yugoslavian president Milosevic and his Serbian forces on the other. Since NATO was bombing Yugoslavia to support Kosovo's ethnic Albanians, we should include Albanians on the NATO side. Russia and China were on the Serbian side, although neither country provided Milosevic with any military support.

The Rules

There are a variety of rules that govern international conflicts, for example, the United Nations Charter and various protocols that arose from the Geneva Convention regarding the safety of civilian populations. We will discuss some of these rules shortly.

Payoffs

Unlike roulette, chess, or Prisoner's Dilemma, which have clearly specified structures and payoffs, the Yugoslavian conflict was neither a gambling game nor a parlor game nor a mathematical abstraction. In a complex situation such as this, it is instructive to look at the players' goals.

NATO's goal, or what U.S. State Department spokesman James Rubin called "political objectives," was "that ethnic Albanian refugees be allowed to return home; that Serb forces withdraw from Kosovo; and that an international peacekeeping force with NATO at its core be installed in Kosovo."

Milosevic's goal was simple: Resist the bombing and hope that NATO would go away and let him rule Yugoslavia in the way to which he had grown accustomed. This may have included various forms of oppressing ethnic Albanians, including but not limited to forced expulsion of Albanians from Yugoslavia, murder, and elimination of basic rights.

What Type of Game Was the Yugoslavian Conflict?

We have examined a variety of game structures. Let's see if we can find one that provides a good model for the Yugoslavian conflict.

Was it like the generals' war game, with NATO acting as General A and Yugoslavia as General B? In Chapter 9 Generals A and B played a ritual, zero-sum war game that was represented by the following payoff matrix:

		General B	
		surrender	resist
	invade	6	5
General A	bomb	7	4
	diplomacy	9	−10

Recall that the payoffs for this game, although somewhat arbitrary, arose from the following considerations:

- If General B resists and General A chooses diplomacy, nothing much will happen, so the diplomacy payoff should be the best of General B's payoffs if he uses the resistance strategy.

- Still going with the resistance strategy, bombing should be preferable to General B over invasion, because an invasion will probably result in General B's surrender, whereas it's easier to resist bombing.

- If General B decides to surrender, General A's best choice is diplomacy, since he would be accomplishing his goal with no casualties or other costs of war. Still considering his cost and casualties, if General B surrenders, General A would prefer bombing to invasion.

Using this game as a guideline, NATO's optimal strategy would have been to invade, Yugoslavia's to resist.

Although the generals' simple war game is similar in some ways to the Yugoslavian conflict, there are major differences. For example, the generals' game doesn't consider international law, which frowns on a nation or alliance of nations invading a sovereign state. It also doesn't consider public opinion. A NATO invasion would have meant casualties for the NATO forces. This might have caused a drop in public support in NATO countries for the conflict.

As we know, NATO chose to bomb, and Yugoslavia chose to resist, at least at the outset. The conflict lasted seventy-eight days, so this was not a one-move game. This is a key difference from the generals' game: Although Milosevic initially resisted, eventually Yugoslavia was overpowered by the bombardment and Milosevic was essentially forced to surrender. NATO "won" this many-move game in the sense that Yugoslavia capitulated, however, great costs were incurred by both sides.

Was the Yugoslavian conflict more like a hostage-rescue situation than a war? Because of the overwhelming superiority of the NATO forces, it could have been thought of this way. In this case Milosevic and his forces were like a terrorist group, holding ethnic Albanians hostage, forcing some to abandon their homes and brutalizing others. Using this model, what should NATO have done?

The primary goal of hostage rescue is to protect the lives of the hostages and other innocent people. The NATO bombing strategy seemed to be at odds with this goal, as it assured the deaths of numerous Yugoslavian civilians as well as many of the "hostages" themselves. In fact, the NATO bombing strategy seemed designed to do whatever was necessary to protect the lives of the rescuers rather than the hostages, certainly not a conventional hostage-rescue strategy.

No SWAT team has ever bombed the region where terrorists are holding hostages. Acceptable strategies call for the equivalent of an invasion to selectively stop terrorists if negotiations break down. In the Yugoslavian conflict not much time was spent negotiating. The result of the NATO strategy was partially successful, since the hostages who were still alive were freed when Milosevic surrendered, however, much of Kosovo was bombed into rubble, leaving the hostages with a bleak sort of freedom. Meanwhile, the "terrorists" retreated from Kosovo to a safe location. From the hostage-rescue perspective, this could hardly be called a victory.

Was the Yugoslavian conflict like a video game? Bombing Yugoslavia was like playing a video game, as follows: Everything was computer-controlled, the pilots saw their targets on a video screen, and the sights and sounds of bombs as they hit the targets (and anything else in the general vicinity) resembled the special effects of video games. Even public opinion was consistent with the video-game model. Civilian casualties caused by NATO bombing didn't arouse much sympathy in NATO countries. To many people, the lives of Yugoslavians appeared to be no different from that of the aliens in video games, existing solely for game players to keep score.

This suggests an alternative to the bombing strategy: NATO could have staged a video-game tournament between the opposing countries. In this case, if NATO won, Milosevic would have had to withdraw all his forces from Kosovo. If the Serbian troops won, NATO would have had to let Milosevic preside over Yugoslavia however he wanted. The skills would have been the same as bombing, nobody would have been hurt, nothing would have been bombed into rubble, and it's unlikely that Milosevic could have found any Serbian video-game players who could have beaten the NATO pilots.

Was the Yugoslavian conflict like Prisoner's Dilemma? When we analyzed Prisoner's Dilemma, we saw the effectiveness of tit for tat as a strategy that was cooperative yet couldn't be exploited. Since NATO's objective was to make Kosovo safe for ethnic Albanians, it seemed that there might have been room for cooperation. Maybe the NATO bombing was just the nasty side of tit for tat.

Upon closer inspection, we find major differences between Prisoner's Dilemma and the Yugoslavian conflict:

Prisoner's Dilemma is a symmetric game, in which the same options are available to each player. In the Yugoslavian conflict, President Milosevic's army was only capable of local carnage, while NATO forces were the most powerful on earth.

In Prisoner's Dilemma there are only two moves, cooperation and nastiness. In Yugoslavia each side had a range of choices.

In Prisoner's Dilemma there must be a clear definition of what it means to be cooperative and what it means to be nasty. In the Yugoslavian conflict, although there was no ambiguity about the nastiness of bombing or of ethnic cleansing, NATO's version of cooperation was what Milosevic called surrender.

It is the ambiguity of *cooperation* that wrecks Prisoner's Dilemma as a model for the Yugoslavian conflict. In fact, the continued bombing made it difficult for Yugoslavians to accept any NATO peace deal. In a *Washington Post* article about a Yugoslavian factory bombed by NATO, Michael Dobbs observed, "Having already lost their jobs and their livelihoods, the workers at the factory would seem to have little left to lose by further resistance to NATO, and therefore little incentive to support a peace deal that would create what amounts to an international protectorate for Kosovo, which Serbs regard as the cradle of their civilization."

Although Yugoslavia finally capitulated, in the eyes of some, Serbian "cooperation" with NATO did not provide a higher reward than being nasty. This violated the payoff structure of Prisoner's Dilemma. The game was zero-sum.

Mixed Strategies and Strategic Planning

When we discussed zero-sum games, we saw that in some situations optimal strategies required the use of chance to make the opponent uncertain which action the player would choose. In the Yugoslavian conflict, a massing of NATO troops along the Kosovo border before the bombing started might have made NATO's strategy unclear to Milosevic. Even if this was a bluff, Milosevic might have opted for a peace plan before any blood was shed, because he may have been unwilling to risk a possible NATO invasion. As it turned out, in apparent ignorance of the principles of zero-sum game theory, President Clinton publicly announced that he would not use ground troops, perhaps delaying Milosevic's eventual surrender.

In another apparent example of the lack of strategic planning by the United States, Richard Holbrooke, then Special U.S. Envoy to the Balkans and Ambassador Designate to the United Nations, made the following, bizarrely Machiavellian comment in an April 22 speech to the Overseas Press Club: "I do not think it is wise or prudent for anyone in the government or out of the government to project how long the bombing will continue. . . . Projections only come back to haunt you." In fact, it took 78 days of bombing, thousands of Yugoslavian civilian casualties, billions of dollars in damage to the Yugoslavian infrastructure, and billions of dollars of costs incurred by the NATO forces before the war ended.

Civilian Casualties and Lottery Winners

Recall from Chapter 1 Maria Grasso cashing in on the $197 million Big Game lottery jackpot, the largest lottery prize ever won by an individual. The odds of winning that jackpot were 1 in 76 million. It was an incredible, lucky accident that Grasso in particular won the jackpot, but it wasn't surprising that someone won.

In the Yugoslavian conflict, NATO was able to avoid any casualties to its pilots; however, its "safe" (to NATO) bombing strategy caused thousands of Yugoslavian civilian casualties. But deliberately causing civilian casualties is a violation of international law. What happened?

On April 12 a NATO missile slammed into a passenger train on a bridge in southern Serbia, leaving at least ten dead and injuring more than a dozen passengers. According to General Wesley Clark, NATO's supreme commander in Europe, the missile was fired at the bridge, not the train. The fact that the train moved onto the bridge at the instant when the bomb reached its destination was, according to Clark, "an uncanny accident." Another uncanny accident occurred three weeks later on May 1 when, according to *Los Angeles Times* reporter Paul Watson, "A NATO air strike blew a civilian bus in half on a bridge in Kosovo, killing between 34 and

60 people, including many children." Watson quotes NATO officials as saying, "Unfortunately after weapon release, a bus crossed the bridge." In another article, Watson reported that "NATO bombers scored several direct hits here in Kosovo's capital yesterday-including a graveyard, a bus station, and a children's basketball court. The targets weren't mentioned when General Wesley Clark, NATO's supreme commander, briefed reporters in Brussels on the air campaign's success."

As reported in a May 31 *Los Angeles Times* article, here is part of an Associated Press summary of other NATO bombing accidents:

> April 5—An attack on a residential area in the mining town of Aleksinac kills 17 people. April 14—75 ethnic Albanian refugees die in an attack on a convoy near Djakovica. April 27—A missile strike in the Serbian town of Surdulica kills at least 20 civilians. May 7—A cluster bomb attack damages a marketplace and the grounds of a hospital in Nis, killing at least 15. May 8—Fighter pilots using outdated maps attack the Chinese Embassy in Belgrade, killing three journalists and injuring 20 other people. May 13—87 ethnic Albanian refugees are killed and more than 100 injured in a late-night NATO bombing of a Kosovo village, Korisa. May 20—At least three people are killed when NATO missiles hit a hospital near a military barracks in Belgrade. May 21—NATO bombs a Kosovo jail, killing at least 19 people and injuring scores. May 21—One Kosovo Liberation Army guerrilla is killed and at least 15 injured in an attack on a stronghold of the rebel force. May 31—NATO missiles slam into a bridge crowded with market-goers and cars in central Serbia, killing at least nine people and wounding 28. May 31—In southern Kosovo, a NATO missile attack occurs near a convoy of journalists, killing a local driver and wounding three people.

In mid-May NATO spokesman Jamie Shea defended the alliance's bombing accuracy, claiming that since the bombing had begun, NATO had dropped about nine thousand missiles and bombs and that only twelve had "gone astray." Shea said, "If you do a mathematical computation, you're talking about a fraction of one percent." Shea didn't mention that the fraction of 1 percent of bombs that "went astray" killed many civilians and caused billions of dollars in damage to the country's infrastructure.

If not deliberate, the bombing of hospitals, buses, trains, residential complexes, marketplaces, embassies, and so on, occurred for the same reason that Maria Grasso won the Big Game: Given enough opportunity, unlikely events become statistical certainties. When enough people buy lottery tickets, there are jackpot winners, and when there is enough bombing of populated areas, civilian targets will be hit.

Long-Term Consequences of the Bombing Strategy

In Prisoner's Dilemma the optimal strategy for the one-move game is quite different from a good strategy in the many-move game. In the many-move game, where players interact for a lengthy period, there is reason for cooperation. It is unclear whether NATO analyzed the long-term consequences of its bombing strategy. For example, although Milosevic surrendered, clearly a short-term victory for NATO, the duration and intensity of the bombing strategy may have the long-term effect of destabilizing the region and causing animosities between NATO and non-NATO countries throughout the world.

A CNN article about the NATO bombing of the Chinese Embassy in Belgrade quoted Japanese prime minister Keizo Obuchi as saying, "It is deplorable that a diplomatic establishment which is supposed to have extraterritorial jurisdiction—moreover an embassy—has been mistakenly hit." Indian foreign minister Jawant Singh said, "Any damage to a diplomatic establishment, intended or otherwise, is to be entirely deplored. This incident along with continuing loss of innocent lives and other untoward consequences only confirms that the very fundamentals of NATO's new approach are wrong." The same article quoted the South African government as saying that "the bombing raids would not only exacerbate the humanitarian tragedy but that they would lead to unfortunate incidents such as the bombing of the Chinese Embassy and others."

In a *San Francisco Chronicle* article, called "Kosovo's Ripple Effect," Ruth Rosen commented, "As a new generation of [Chinese] students watched the United States bomb both Kosovo and Belgrade—without declaring war, without the consent of the American peoples' representatives and without a mandate from the United Nations—democracy began to seem like just another nine-letter word." Rosen further noted, "Who wins and loses a war is not always so clear. For the Chinese Communist Party, the war in Kosovo was an unintended blessing. . . . In this case, among the many casualties of the war in Kosovo is one generation's belief in democracy and human rights."

Another long-term consequence of the bombing strategy is its effect on the Yugoslavian economy. President Clinton described Milosevic as a leader who would "rather rule over rubble than not rule at all." An Associated Press article quoted General Klaus Naumann, head of NATO's military arm, as saying, "We may have one flaw in our thinking: We believe that no responsible man who is at the helm of a country like Yugoslavia can wish to run the risk that his entire country will be bombed into rubble before he . . . complies with the demands of the international community."

It is unclear why NATO put itself in the position of returning a million ethnic Albanian refugees to a region that had been reduced, at least partially, to rubble.

In a *Washington Post* article titled "War Leaves Yugoslav Economy a Shambles," reporter Michael Dobbs observed, "Yugoslav officials say that the damage from NATO bombs has reached the $100 billion mark. By some estimates the bombing has set Yugoslavia back one or even two decades." Dobbs went on to note, "With the destruction of its largest factories by NATO bombs, unemployment is escalating and prospects for economic reconstruction seem bleak."

In a *Los Angeles Times* article, entitled "NATO Bombing Left Serbian City in Toxic Nightmare," Mark Fineman noted that when NATO bombs demolished a refinery, a fertilizer plant, and an American-built petrochemical complex in Pancevo, Yugoslavia, they "released a toxic cloud so dense and potentially lethal that its effects can be felt here even today—and will be, perhaps, for decades to come." It seems that perhaps inadequate attention was paid to this negative payoff resulting from the bombing strategy.

Rules of the Game

All games, from chess to professional football, have specific rules, with referees to enforce them. In the United States, legislative bodies write laws that regulate our daily routines. There are various international rules that are supposed to regulate wars and other conflicts. In the Yugoslavian conflict, Milosevic's policies of ethnic cleansing, including murder and forced relocation of ethnic Albanians, were in clear violation of international law. As a result of these policies, Milosevic was declared a war criminal by an international war crimes tribunal. Milosevic, however, may not be the only one who violated international law.

When NATO bombs struck the Chinese Embassy in Belgrade, Chinese U.N. ambassador Qin Huasun told the U.N. Security Council that NATO had committed a "crime of war." Vietnam's Foreign Ministry called the bombing of the embassy a "serious violation of international law." A CNN article quoted U.N. human rights chief Mary Robinson as saying, "In the NATO bombing of the Federal Republic of Yugoslavia, large numbers of civilians have been incontestably killed, civilian installations targeted on the basis that they are or could be of military application, and NATO remains the sole judge of what is or is not acceptable to bomb." Robinson appealed to the U.N. Security Council "to have a say on whether a prolonged bombing campaign was in fact legal under the U.N. charter."

Article 52, section 4, of the Protocol to the Geneva Conventions of August 12, 1949, relating to the "Protection of Victims of International Armed Conflicts (Protocol I)" (entered into force December 7, 1978), states, "Indiscriminate attacks are prohibited." This includes "those which employ a method or means of combat which cannot be directed at a specific military objective" as well as "an attack by bombardment by any methods or means which treats as a single military objective a number of clearly separated and distinct military objectives located in a city, town, village or other area containing a similar concentration of civilians or civilian objects." And also "an attack which may be expected to cause incidental loss of civilian life, injury to civilians, damage to civilian objects, or a combination thereof, which would be excessive in relation to the concrete and direct military advantage anticipated."

In a *New York Times* article, reporter Michael Gordon wondered "whether it is really possible to dependably distinguish civilian from military targets at the altitude at which NATO planes customarily operate, 15,000 feet."

In a May 31 *New York Times* article, Eric Schmitt noted, "Increasing attacks on civilian infrastructure, like electric power grids that operate hospitals and water-pumping stations, as well as civilian casualties, have prompted some NATO allies like Italy and Greece to call for a moratorium in the air campaign, as an incentive for Yugoslav President Slobodan Milosevic to meet NATO's demands."

In a *New York Times* column, Thomas L. Friedman stated, "As the Pentagon will tell you, air power alone brought this war to a close in 78 days for one reason—not because NATO made life impossible for the Serb troops in Kosovo (look how much armor they drove out of there), but because NATO made life miserable for the Serb civilians in Belgrade. . . . Once NATO turned out the lights in Belgrade, and shut down the power grids and the economy, Belgrade's citizens demanded an end to the war. It's that simple."

Conclusion

The conflict in Yugoslavia had similarities to a variety of games but couldn't be effectively modeled by any particular one. The NATO bombing strategy, despite its short-term success, may have many negative, long-term consequences. Although it is true that Milosevic surrendered, the bombing strategy caused thousands of civilian casualties and massive destruction of Yugoslavia's civilian infrastructure. This possibly could have been avoided had NATO used another strategy, either military or diplomatic. In addition to its devastating effects on the people of Yugoslavia, the long-term consequences of the NATO bombing strategy may include a breakdown of trust between NATO and non-NATO countries.

From a game theoretic point of view, the violations of rules governing international conflicts, by both Milosevic and by NATO, were perhaps the most disturbing part of the Yugoslavian conflict. Civilized countries pride themselves on being dedicated to the rules of law. Regardless of whether a particular cause is just, if a ruthless leader, a country, or a powerful alliance of countries disregards international law, the world can become a vicious arena of uncontrollable violence, a place where the cooperative paradigm suggested by Prisoner's Dilemma has little chance of ever taking hold.

Bibliography

Axelrod, Robert. *The Evolution of Cooperation*, Basic Books, 1984.

Bennett, Deborah J. *Randomness*, Harvard University Press, 1998.

Casti, John L. *Searching for Certainty*, Morrow, 1990.

Davis, Morton. *Game Theory*, Basic Books, 1970.

Diaconis, Percy, and Frederick Mosteller. *Methods for Studying Coincidences*, Journal of the American Statistical Association, Vol. 84, 1989.

Feller, William. *An Introduction to Probability Theory and Its Applications*, Volume 1, Wiley, 1950.

Gleick, James. *Chaos*, Penguin, 1987.

Greene, Brian. *The Elegant Universe*, W.W. Norton and Co., 1999.

Musashi, Miyamoto. *A Book of Five Rings*, Overlook Press, 1974.

Orkin, Mike. *Can You Win?*, W.H. Freeman and Co., 1991.

Penrose, Roger. *The Emperor's New Mind*, Penguin, 1989.

Peterson, Ivars. *The Jungles of Randomness*, Wiley, 1998.

Savage, Leonard J. *The Foundations of Statistics*, Wiley, 1954.

Sorrell, Roderic, and Amy Sorrell. *The I Ching Made Easy*, Harper, San Francisco, 1994.

Thorp, Edward O. *The Mathematics of Gambling*, Gambling Times, 1984.

Thorp, Edward O. *Beat the Dealer*, Vintage Books, 1962.

Wilhelm, Richard, and Cory Baynes. *The I Ching*, Princeton University Press, 1977.

Wong, Stanford. *Professional Video Poker*, Pi Yee Press, 1991.

Index